Advances in Missile Guidance Theory

Advances in Missile Guidance Theory

Joseph Z. Ben-Asher and Isaac Yaesh
Israel Military Industries—Advanced Systems Division
Ramat-Ha'Sharon, Israel

Volume 180
PROGRESS IN
ASTRONAUTICS AND AERONAUTICS

Paul Zarchan, Editor-in-Chief
Charles Stark Draper Laboratory, Inc.
Cambridge, Massachusetts

Published by the
American Institute of Aeronautics and Astronautics, Inc.
1801 Alexander Bell Drive, Reston, Virginia 20191-4344

5 4 3 2 1

ISBN 1-56347-275-9

Table of Contents

Preface

This publication deals with advanced guidance methods and is oriented toward practicing engineers or engineering students. During the past few decades, ideas taken from linear quadratic (LQ) theories have been successfully applied to guidance problems. In recent years, LQ optimization methods have become widely used in various control and estimation problems that have subsequently opened up new possibilities in the guidance field. The purpose of this book is to explore some new applications of LQ theories to the optimal guidance problem, not only focusing on the mathematical results but also imparting physical insight.

New guidance laws that are based on LQ one-sided optimization theory and on differential game theory are introduced. In addition, old guidance laws are rederived and set in a more general framework. The main theme is to systematically analyze guidance problems with increasing complexity. The guidance problems are formulated and solved with state-space representations in a rather informal mathematical style. These problems, combined with numerous examples, make this publication a useful guide for the missile guidance engineer. Its style and the fact that it is self-contained (most of the theory needed is given in Chapter 2) makes it suitable as a textbook for advanced courses on missile guidance or as a reference for practicing engineers.

Throughout the monograph a special effort was made to give analytical (closed-form) solutions to the optimization problems. For this purpose, we used mostly MAPLE™ programs, the listings of which are incorporated in the text. For the numerical examples we used MATLAB™ programs, the relevant subroutines of which are also part of the text. A diskette with the source code listings of both the MAPLE and the MATLAB subroutines is included, so that the interested reader can duplicate and extend our results. The numerical examples given in the book usually involve a comparison of several guidance laws with the aim of helping the reader to understand which guidance law is suitable for each guidance problem.

Although in several chapters some results from existing theories such as H_∞ theory can be invoked, we prefer making our own derivations, so that we do not burden the readers by sending them to other references on LQ methods.

In Chapter 1 the various objectives of the book and the motivation for their choice are restated. A short historical background is given to orient this research effort in the area of missile guidance. Finally, the mathematical modeling employed throughout the book and a general formulation of the problems dealt with are presented.

Chapter 2 develops the mathematical tools needed for the derivation of LQ guidance methods. It contains a review of the theory of LQ control, presents necessary and sufficient conditions for optimality, and describes the idea of performance evaluation by adjoint analysis in state space.

LQ guidance methods based on the one-sided optimization problem—the classical optimal control problem—are derived in Chapter 3. Optimal guidance of an ideal missile against a nonmaneuvering target with and without terminal velocity constraints is studied. The effects of maneuvering targets, nonideal pursuer dynamics, and nonideal evader dynamics are discussed. Existing guidance schemes, such as proportional navigation and advanced proportional navigation, are rederived. A numerical study is carried out to compare the various methods.

Chapter 4 treats LQ guidance methods based on the two-sided optimization problem—the differential game problem. It is developed in similar fashion to the preceding chapter. First, the optimal guidance law of an ideal missile against an ideal target, with and without terminal velocity constraints is developed. Then, effects of target maneuvers and of nonideal dynamics are studied. Finally, numerical examples that compare the various methods are presented.

In Chapter 5 guidance methods with improved performance in terms of sensitivity reduction are discussed. This task is carried out by weighting velocity and acceleration terminal values in the cost to reduce the sensitivity to "time-to-go" errors. The reduction in sensitivity is demonstrated by numerical examples in which we introduce such errors into various guidance schemes and compare performance in terms of miss distance and control effort.

Chapter 6 extends the idea of sensitivity reduction by considering parameter uncertainties in the system. Robust design with respect to system dynamics is developed based on H_∞ theory. Although the theory in all previous chapters could be presented as special cases of the H_∞ theory, the theory in this chapter could be obtained only from recent results in the field of H_∞. A comparison of robust and standard guidance laws is made.

The problem of intercepting multiple targets is formulated and solved in Chapter 7. To this end, LQ methods are employed with terminal and intermediate point constraints (the required extension of the theory is given in this chapter) and closed-form feedback schemes against stationary targets are obtained.

Because of the extensive use of MAPLE as a tool for obtaining analytical results, a short introduction to this symbolic code and how it is used in this book is given in Appendix B. A description of the MATLAB package used for the numerical simulations is given in Appendix C.

We believe that the book is a natural extension to Zarchan's excellent book *Tactical and Strategic Missile Guidance* (Vol. 176 of the Progress in Astronautics and Aeronautics series), especially Chapter 8, Advanced Guidance Laws. Paul Zarchan contributed to our book by making many useful comments and recommendations and encouraging us to complete the task. We are very grateful for his useful advice and comments. We are also indebted to Aron Pila for reading the manuscript, making numerous remarks, and improving it both in form and content. Finally, the authors would like the emphasize the fact that the advanced guidance laws of the book may be used in a variety of aeronautical and astronautical applications where optimal interception and rendezvous can be just as efficient in directing a vehicle to its destination (e.g., landing on an asteroid) as classical methods. While the applications of the advanced guidance methods might be diverse and some more lethal than others, we maintain that science by itself is never good or bad, only its use can be characterized as such.

Joseph Z. Ben-Asher
Isaac Yaesh
May 1998

Chapter 1

Introduction

I. Objectives and Motivation

THIS book deals with advanced missile guidance methods. More specifically, since the terminology in this field is ambiguous, it is about the terminal guidance of intercepting missiles, where it is assumed that the trajectory for this flight phase can be linearized around some nominal collision course. The main objective of this book is the exploration of guidance strategies that will perform the required interception while optimizing a certain performance index, where the latter is taken to be of quadratic form. Hence, we call the ensuing methods linear quadratic methods.

In recent years, linear quadratic (LQ) theories have matured and, with the availability of appropriate software design tools, have come to be widely employed. While in the past several ideas based on the LQ approach were successfully applied to guidance problems, many new possibilities continue to appear, particularly those based on recent advances in LQ theory that have not yet been explored. These unexplored possibilities provide the motivation for our research. With the aim of pursuing new approaches for the application of LQ theories to optimal guidance, we will cover methods based on one-sided optimization problems and on differential game problems. While proposing new approaches, many of the classical methods will be presented as well, typically as special cases of some more general formulations.

II. Historical Background

Since the first successful test of the Lark missile in December 1950, proportional navigation (PN) has come to be widely employed by homing missiles.[1] Under this scheme the missile is governed by

$$n_c = N' V_c \dot{\lambda} \tag{1.1}$$

where n_c is the missile's normal acceleration, $\dot{\lambda}$ the line-of-sight (LOS) rate, V_c the closing velocity, and N' the so-called navigation constant ($N' > 2$).

In the mid-1960s[2] it was realized that PN with $N' = 3$ is, in fact, an optimal strategy for the linearized problem when the following cost is minimized (control effort),

$$J = \int_0^{t_f} n_c^2(t)\, \mathrm{d}t \tag{1.2}$$

subject to $y(t_f) = 0$, where y is the relative interceptor-target separation and t_f is the nominal collision time. For this case $\dot{\lambda}$ may be expressed as

$$\dot{\lambda} = \frac{y + \dot{y}t_{go}}{V_c(t_{go})^2} \tag{1.3}$$

and hence

$$n_c = \frac{3(y + \dot{y}t_{go})}{t_{go}^2} \tag{1.4}$$

Additional PN constants have resulted when the average separation has been penalized quadratically in addition to the control effort.[3]

In cases where target maneuvers are significant, extensions of the PN law have been developed, such as augmented proportional navigation (APN),[4] where the commanded interceptor's acceleration n_c depends on the target acceleration n_T, so that

$$n_c = \frac{N'(y + \dot{y}t_{go})}{t_{go}^2} + \frac{N'}{2}n_T \tag{1.5}$$

It was realized that when the transfer function relating the commanded and actual accelerations n_c and n_L, respectively, has a significant time lag (with respect to the final time t_f), the APN law can lead to a significant miss distance. To overcome this difficulty, the minimum effort [guidance] law (MEL) was proposed whereby J [Eq. (1.2)] is minimized, subject to state equation constraints that include the missile's dynamics of the form

$$\frac{n_L}{n_c} = \frac{1}{1 + Ts} \tag{1.6}$$

where s is the differentiation operator, namely, $s = \mathrm{d}/\mathrm{d}t$, and T the missile's time lag.

The resulting guidance law easily overcomes the large time lag problem, but is strongly dependent on the time constant and the time-to-go, namely

$$n_c = \frac{N'(y + \dot{y}t_{go})}{t_{go}^2} + \frac{N'}{2}n_T - n_L K_L \tag{1.7}$$

where t_{go} is the time-to-go (namely $t_f - t$), and

$$N' = \frac{6h^2(e^{-h} - 1 + h)}{2h^3 + 3 + 6h - 6h^2 - 12he^{-h} - 3e^{-2h}} \tag{1.8}$$

$$K_L = \frac{T^2}{t_{go}^2}(e^{-h} + h - 1) \tag{1.9}$$

and where $h = t_{go}/T$.

Another approach to pursuit evasion in the LQ context was proposed by Baron.[5] He formulated the problem as a differential game where the target is allowed to maneuver to maximize the miss distance and the pursuer's control effort while minimizing its own control effort, and simultaneously the pursuer tries to minimize the same cost function, thus forming a "zero sum" game. A zero miss can be guaranteed if the pursuer is more maneuverable. The strategy for this case will be referred to as the "game-theoretic optimal intercept." The result is still of the form (1.1) but the PN gain is now

$$N' = \frac{3}{1-\gamma^{-2}} \tag{1.10}$$

where γ represents the relative maneuverability.

Notice that this approach removes the need for estimating the opponent's acceleration but typically requires more acceleration as a result of its being a worst-case design. (The target is assumed to maneuver optimally.)

Apart from the previously mentioned LQ approaches, there have been other attempts made to cast the pursuit-evasion problem into an optimization problem. In particular, important results have been obtained with hard-bounded control limits rather than the soft limits imposed by quadratic penalties.[6,7] Some research effort has gone into investigating nonlinear pursuit-evasion games in the spirit of R. Isaacs[8]—the founder of differential game theory (see Ref. 9), while others introduced only the nonlinear effect of control saturation in an otherwise linear formulation.[10] The issues of target maneuvers estimation and stochastic optimization of guidance laws have also been research topics in the past decades.[11]

In this book, however, we confine ourselves to purely LQ methods. The approach taken is deterministic—although issues of sensitivity reduction and robust guidance form an important part of this work—and the models are all linear with the control being softly bounded by the quadratic cost.

III. Mathematical Modeling

At the outset we shall make the following assumptions:

1) The pursuit-evasion conflict is two-dimensional in the horizontal plane.

2) The speeds of the pursuer p and the evader (the target) e are constant.

3) The trajectories of p and e can be linearized around their collision course.

The geometry of the planar interception scenario around a collision triangle is depicted in Fig. 1.1.

We assume that the collision condition is satisfied, namely,

$$V_p \sin\left(\gamma_{p_0}\right) - V_e \sin\left(\gamma_{e_0}\right) = 0 \tag{1.11}$$

where V_e and V_p are the pursuer's and evader's velocities, and γ_{p_0} and γ_{e_0}, the pursuer's and evader's nominal heading angles, respectively. In this case, the nominal closing velocity V_c is given by

$$V_c = -\dot{R} = V_p \cos\left(\gamma_{p_0}\right) - V_e \cos\left(\gamma_{e_0}\right) \tag{1.12}$$

and the nominal terminal time is given by

$$t_f = t_0 + \frac{R(t_0)}{V_c} \tag{1.13}$$

where $R(t)$ is the nominal length of the LOS at time t and t_0 is the initial time.

If we allow Y_e and Y_p to be the separation of the pursuer and the evader, respectively, from the nominal LOS, and let y be the relative separation (namely $y \equiv Y_e - Y_p$), we obtain the following dynamic equation:

$$\dot{y} = \dot{Y}_e - \dot{Y}_p = V_e \sin\left(\gamma_{e_0} + \gamma_e\right) - V_p \sin\left(\gamma_{p_0} + \gamma_p\right) \tag{1.14}$$

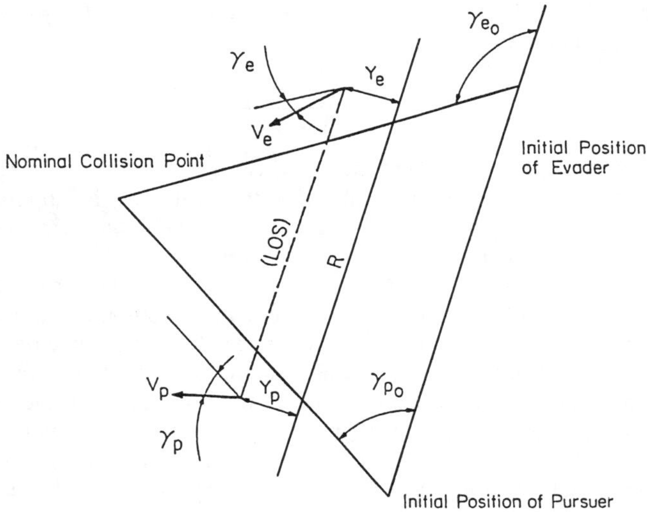

Fig. 1.1 Two-dimensional pursuer-evader engagement geometry.

where γ_p and γ_e are the deviations of the pursuer's and evader's headings, respectively, as a result of control actions applied (see Fig. 1.1). If these deviations are small enough, we may use small angles approximation to obtain

$$\sin\left(\gamma_{p_0}+\gamma_p\right)\approx\sin\left(\gamma_{p_0}\right)+\cos\left(\gamma_{p_0}\right)\gamma_p \tag{1.15}$$

$$\sin\left(\gamma_{e_0}+\gamma_e\right)\approx\sin\left(\gamma_{e_0}\right)+\cos\left(\gamma_{e_0}\right)\gamma_e \tag{1.16}$$

Substituting the results into Eqs. (1.11) and (1.14) we find [using Eq. (1.11)] that $\dot{y}$ becomes

$$\dot{y}=\dot{Y}_e-\dot{Y}_p=V_e\cos\left(\gamma_{e_0}\right)\gamma_e-V_p\cos\left(\gamma_{p_0}\right)\gamma_p \tag{1.17}$$

We can also find an expression for the LOS angle and its rate of change. Let λ be the LOS angle, and without loss of generality let $\lambda(t_0)=0$.

If we denote $\lambda(t)$ as

$$\lambda(t)=y/R \tag{1.18}$$

then

$$\dot{\lambda}(t)=\frac{d}{dt}\left(\frac{y}{R}\right)=\frac{y}{V_c(t_f-t)^2}+\frac{\dot{y}}{V_c(t_f-t)} \tag{1.19}$$

and thus we have obtained Eq. (1.3).

Allowing the controls to be the normal accelerations and defining

$$x_1\equiv y \qquad x_2\equiv\dot{y} \qquad w\equiv V_e\cos\left(\gamma_{e_0}\right)\dot{\gamma}_e$$
$$u\equiv -V_p\cos\left(\gamma_{p_0}\right)\dot{\gamma}_p \tag{1.20}$$

we obtain the following state-space formulation by differentiating Eq. (1.17):

$$\dot{x}_1 = x_2 \qquad \dot{x}_2 = u + w \tag{1.21}$$

or in matrix form

$$\dot{x} = Ax + Bu + Dw$$

$$A = \begin{bmatrix} 0 & 1 \\ 0 & 0 \end{bmatrix} \qquad B = \begin{bmatrix} 0 \\ 1 \end{bmatrix} \qquad D = \begin{bmatrix} 0 \\ 1 \end{bmatrix} \tag{1.22}$$

If both the pursuer and the evader have direct control of their normal accelerations, we shall refer to them as "ideal" and Eq. (1.22) will be the governing equation.

If we assume a first-order lag with time constant T for the pursuer's dynamics and no evader dynamics, we then have the following equations in state-space form:

$$\dot{x} = Ax + Bu + Dw$$

$$A = \begin{bmatrix} 0 & 1 & 0 \\ 0 & 0 & 1 \\ 0 & 0 & -\dfrac{1}{T} \end{bmatrix} \qquad B = \begin{bmatrix} 0 \\ 0 \\ \dfrac{1}{T} \end{bmatrix} \qquad D = \begin{bmatrix} 0 \\ 1 \\ 0 \end{bmatrix} \tag{1.23}$$

where the control u is now the pursuer's commanded acceleration.

Finally, when both pursuer and evader are nonideal, we find that

$$\dot{x} = Ax + Bu + Dw$$

$$A = \begin{bmatrix} 0 & 1 & 0 & 0 \\ 0 & 0 & 1 & 1 \\ 0 & 0 & -\dfrac{1}{T} & 0 \\ 0 & 0 & 0 & -\dfrac{1}{\Theta} \end{bmatrix} \qquad B = \begin{bmatrix} 0 \\ 0 \\ \dfrac{1}{T} \\ 0 \end{bmatrix} \qquad D = \begin{bmatrix} 0 \\ 0 \\ 0 \\ \dfrac{1}{\Theta} \end{bmatrix} \tag{1.24}$$

where Θ is the time constant representing the evader's acceleration control system and w is its acceleration command regarded as a disturbance.

In this book we will solve various guidance problems that relate to Eqs. (1.22–1.24). We will always assume that the pursuer has perfect knowledge on $x(t)$ at time t. Thus, the dependency of u on x will be of state-feedback form. This is the form obtained from solutions of the associated one-sided LQ optimization problems (such as the previously described PN). This formulation may also include cases when $w(t)$ is known to the pursuer. In such cases the pursuer's actions will include an additional feed-forward term that uses this knowledge—similarly to the APN scheme.

When $w(t)$ is unknown but of finite energy (i.e., $\int_{t_0}^{t_f} w^2(t)\,dt < \infty$), we will end up solving a two-sided LQ optimization problem, namely an LQ differential game. Again, the solution will become state feedback. This formulation will be extended to deal with sensitivity issues and system uncertainties, thus yielding what we may call robust guidance methods.

References

[1]Zarchan, P., *Tactical and Strategic Missile Guidance*, Vol. 124, Progress in Astronautics and Aeronautics, AIAA, Washington, DC, 1990, Chap. 2.

[2]Bryson, A. E., and Ho, Y. C., *Applied Optimal Control*, Hemisphere, New York, 1975, Chap. 5.

[3]Kreindler, E., "Optimality of Proportional Navigation," *AIAA Journal*, Vol. 11, No. 6, 1973, pp. 878–880.

[4]Nesline, W. N., and Zarchan, P., "A New Look at Classical vs Modern Homing Missile Guidance," *Journal of Guidance and Control*, Vol. 4, No. 1, 1981, pp. 78–84.

[5]Ho, Y. C., Bryson, A. E., and Baron, S., "Differential-Games and Optimal Pursuit-Evasion Strategies," *IEEE Transactions Auto. Control*, Vol. C-10, No. 4, 1965, pp. 385–389.

[6]Gutman, S., and Leitman, G., "Optimal Strategies in the Neighborhood of a Collision Course," *AIAA Journal*, Vol. 14, No. 9, 1976, pp. 1210–1212.

[7]Gutman, S. "On Optimal Guidance for Homing Missiles," *Journal of Guidance and Control*, Vol. 2, No. 4, 1979, pp. 296–300.

[8]Isaacs, R., *Differential Games*, Wiley, New York, 1965, Chap. 1.

[9]Green, Shinar, J., and Guelman, M., "Game Optimal Guidance Law Synthesis for Short Range Missiles," *Journal of Guidance, Control, and Dynamics*, Vol. 15, No. 1, 1992, pp. 191–197.

[10]Rusnak, I., and Levi, M., "Optimal Guidance for High-Order and Acceleration Constrained Missile," *Journal of Guidance, Control, and Dynamics*, Vol. 14, No. 3, 1991, pp. 589–596.

[11]Deyst, J., Jr., and Price, G. F., "Optimal Stochastic Guidance Laws for Tactical Missiles," *Journal of Spacecraft and Rockets*, Vol. 10, No. 5, 1973, pp. 301–308.

Chapter 2

Mathematical Background

I. Scope and Preliminaries

SINCE many optimal guidance problems are reducible to special cases of LQ differential games, the design of such algorithms requires some familiarity with game theory. The main purpose of this chapter is to provide the requisite mathematical background for solving various types of LQ differential games stemming from guidance problems. Necessary and sufficient conditions for optimality will be derived and discussed. We also will introduce the reader to performance evaluation techniques based on state-space adjoint analysis.

Consider the following state-space equation:

$$\dot{x} = Ax + B_1 w + B_2 u + B_3 d \tag{2.1}$$

$$x(0) = x_0 \tag{2.2}$$

and the associate cost function:

$$J(u, w) = \int_0^{t_f} (x^T Qx + u^T Ru - \gamma^2 w^T w)\,\mathrm{d}t + x^T(t_f)Qx(t_f) \tag{2.3}$$

We shall use the symbols w, u, and d throughout as outlined in the sequel.

The control signal $u(t)$ defines the pursuer's actions so as to minimize the cost function J of Eq. (2.3). Often, it may be the input to the acceleration control loop of the pursuing missile that will be modeled as a single time lag system (i.e., a first-order linear transfer function) of known or unknown value. If this time lag is negligible, then the control is the actual normal acceleration (see Chapter 1). The control signal will not be hard limited, rather it will be constrained softly by the quadratic penalty on u in Eq. (2.3). The control signal $u(t)$ serves as the minimizer in our optimization problem, and because of the soft constraint will avoid applying excessive actions.

The unknown disturbance signal $w(t)$ is assumed to be of finite energy, namely

$$\int_0^{t_f} w^T w\,\mathrm{d}t < \infty \tag{2.4}$$

It will serve to describe the evasive actions of the target (evader). A smart evader will choose its actions based on perfect measurements of the state vector $x(t)$; that is, the evader's actions $w(t)$ may stem from closed-loop strategies of the form:

$$w(t) = f_e[x(t), t] \tag{2.5}$$

Like the control signal $u(t)$, the disturbance signal $w(t)$ will be softly constrained by the quadratic penalty on its actions in Eq. (2.3) rather than being hard limited. The disturbance signal $w(t)$ serves as the maximizer of J. Because of the soft constraint, the evader will also avoid using excessive actions. As γ in Eq. (2.3) is increased to infinity, the evader will be forced to nullify its actions and the optimization problem then becomes one sided (i.e., a standard LQ control problem).

A known disturbance signal $d(t)$ serves as a deterministic evasive maneuver whose time history for the whole interception duration is known to the pursuer in advance. This type of disturbance is used by nonsophisticated targets (evaders) having the capability of applying preplanned maneuvers such as constant acceleration, sinusoidal acceleration, constant jerk (the time derivative of the acceleration), etc.

Remarks:

1) The case $B_1 = B_3 = D$ may have another practical interpretation. It may be that the target's acceleration $(w + d)$ is being estimated by the pursuer, and thus d is the estimated value and w the estimation error. The problem therefore is to design a guidance scheme that takes into account the worst possible measurement error.

2) In the literature of LQ differential games[1] another type of disturbance is considered that stems from open-loop strategies of the following type:

$$w(t) = f_e'(x_0, t)$$

Our assumption is that if the target is smart enough to compute $w(t) = f_e'(x_0, t)$, then it can also compute $w(t) = f_e[x(t), t]$. Therefore, we do not treat these open-loop strategies, and confine ourselves only to those nonsophisticated targets that can apply preplanned maneuvers or to those that are smart enough to use closed-loop feedback strategies.

The state vector $x(t)$ whose time history describes the conflict in the vicinity of some nominal trajectory (i.e., collision course) shall normally include interceptor target relative position and velocity, and possibly interceptor acceleration, target acceleration, etc. The initial value of the state vector $x(0)$ is assumed to be fixed (often zero) for design purposes and is not used to the advantage of the evader [i.e., the evader has no control over $x(0)$]. The state vector at the end of the conflict $x(t_f)$ is free (i.e., unconstrained).

We assume that the state weighting matrix $Q \geq 0$, where in most applications we will simply take $Q = 0$. The final state weighting matrix is Q_f. In most applications (but not always) we assume $Q_f \geq 0$, and impose the requirement of small miss distance by taking large positive values in the entries of Q_f that correspond to the interceptor target relative distance.

For the control weighting matrix R we assume $R > 0$, and thus deal only with nonsingular LQ games. This assumption makes the "completing-the-square" arguments relatively easy to use and makes possible the direct application of the calculus of variations to obtain candidates for optimal controllers and disturbance signals.

The system's dynamic matrix (see Chapter 1) is A. The matrix B_1 is the unknown disturbance feed-through matrix (disturbance to state derivative). The control feed-through matrix (control to state derivative) is B_2 while B_3 is a deterministic disturbance feed-through matrix (disturbance to state derivative).

Finally, γ is a design parameter that somehow represents the relative maneuverability of the pursuer with respect to the evader.

II. Game Rules Specifications

In proposing the differential game, we have to define exactly what we mean by "game," "equilibrium," and "state feedback." We adopt the terminology of Ref. 1 and assume the following.

1) We limit ourselves to two-person games, where the players are the pursuer (missile, interceptor) and the evader (target).

2) The time interval $[0, t_f]$ is defined a priori and denotes the duration of the game.

3) The information pattern for both players is given by

$$\eta(t) = \{x(s), 0 \leq s \leq t\} \qquad t \in [0, t_f] \tag{2.6}$$

This type of information structure is referred to by Ref. 1 as a "closed-loop perfect-state" pattern. The next claim asserts that it makes a difference who plays first. Denote by $\min_u \max_w J(u, w)$ the situation where u first chooses its strategy; the choice becomes known to w, and then w makes its own choice of strategy. The denotation of $\max_w \min_u J(u, w)$ refers to the opposite order of plays.

Claim (see Ref. 1).

$$\bar{J} \equiv \min_u \max_w J(u, w) \geq \max_w \min_u J(u, w) \equiv \underline{J} \tag{2.7a}$$

Proof: Denote by (u_1, w_1) the solution of $\min_u \max_w J(u, w)$ and by (u_2, w_2) the solution of $\max_w \min_u J(u, w)$. Consider first the $\max_w \min_u J(u, w)$ problem. For this case u minimizes $J(u, w)$ after w has chosen $w = w_2$, therefore for any u

$$J(u_2, w_2) \leq J(u, w_2) \tag{2.7b}$$

Similarly, from the $\min_u \max_w J(u, w)$ problem, for which w maximizes $J(u, w)$ after u has chosen $u = u_1$, we conclude that for any w

$$J(u_1, w_1) \geq J(u_1, w) \tag{2.7c}$$

Substituting $u = u_1$ in Eq. (2.7b) and $w = w_2$ in Eq. (2.7c), we obtain $J(u_2, w_2) \leq J(u_1, w_2)$ and $J(u_1, w_1) \geq J(u_1, w_2)$, respectively, hence $J(u_1, w_1) \geq J(u_2, w_2)$.
QED

Since we have a game in which u tries to minimize J and w tries to minimize $-J$, we say that our game is zero sum. If the upper value of the game $\bar{J} = J(u_1, w_1)$ is equal to the lower value of the game $\underline{J} = J(u_2, w_2)$, we have $u_1 = u_2 = u^*$ and $w_1 = w_2 = w^*$ and

$$J^* = J(u^*, w^*) = \bar{J} = \underline{J} \tag{2.8}$$

Then the pair (u^*, w^*) is called a saddle-point solution and we say that the game admits a saddle point. Such a saddle-point solution will, therefore, satisfy by Eqs. (2.7b) and (2.7c) the following pair of saddle-point inequalities:

$$J(u^*, w) \leq J(u^*, w^*) \leq J(u, w^*) \tag{2.9}$$

where $J(u^*, w^*)$ is called the value of the zero-sum game that we are dealing with. The meaning of the saddle-point inequality Eq. (2.9) is as follows: If the pursuer u chooses to apply a strategy u other than u^*, then the evader can apt for the strategy w^*, thus gaining an advantage that leads to $J(u^*, w^*) \leq J(u, w^*)$. The same goes for the evader; that is, if the evader chooses some $w \neq w^*$, then the pursuer can choose to apply u^*, leading to $J(u^*, w) \leq J(u^*, w^*)$. Therefore, both players should maintain their saddle-point strategies to minimize potential loss (J for the pursuer and $-J$ for the evader).

III. Candidates for Saddle-Point Solutions

In this section we assume the existence of saddle-point solutions u^* and w^* that satisfy Eq. (2.9). By application of the calculus of variations, we can find candidates for u^* and w^* satisfying first-order necessary conditions. Our derivation closely follows along the lines of Ref. 2.

First let x^* be the state trajectory that is associated with u^* and w^*, where the governing state equation is

$$\frac{\mathrm{d}x^*}{\mathrm{d}t} = Ax^* + B_1 w^* + B_2 u^* + B_3 d \qquad x^*(0) = x_0 \tag{2.10}$$

Suppose that $w = w^*$ is fixed and u^* is perturbed to be

$$u = u^* + \eta \bar{u} \tag{2.11}$$

where $\bar{u}$ is the perturbing signal, and η is a small real-valued scalar.

Then x, the perturbed state vector, satisfies

$$\frac{\mathrm{d}x}{\mathrm{d}t} = Ax + B_1 w^* + B_2(u^* + \eta \bar{u}) + B_3 d \qquad x(0) = x_0 \tag{2.12}$$

Subtracting Eq. (2.10) from Eq. (2.12) results in

$$\frac{\mathrm{d}(x - x^*)}{\mathrm{d}t} = A(x - x^*) + B_2 \eta \bar{u} \tag{2.13}$$

Defining $\bar{x}$ as

$$x = x^* + \eta \bar{x} \tag{2.14}$$

we find from Eq. (2.13) that

$$\frac{\mathrm{d}\bar{x}}{\mathrm{d}t} = A\bar{x} + B_2 \bar{u} \qquad \bar{x}(0) = 0 \tag{2.15}$$

Substituting Eqs. (2.11) and (2.14) into J of Eq. (2.3) we have

$$\begin{aligned} J = J_\eta(u, w^*) = &\int_0^{t_f} [(x^* + \eta\bar{x})^T Q (x^* + \eta\bar{x}) + (u^* + \eta u)^T R (u^* + \eta u)]\,\mathrm{d}t \\ &+ \left[x^*(t_f) + \eta\bar{x}(t_f)\right] Q_f \left[x^*(t_f) + \eta\bar{x}(t_f)\right] \\ &- \gamma^2 \int_0^{t_f} w^{*T} w^* \,\mathrm{d}t \end{aligned} \tag{2.16}$$

Since u^* is, by definition, minimizing, $J_\eta(u, w^*)$ cannot be an improvement over $J(u^*, w^*)$; hence, its minimum must be at $\eta = 0$. Consequently, the linear coefficient of η in Eq. (2.16) must also be zero (otherwise, there is always some η that improves J), that is,

$$\int_0^{t_f} [\bar{x}^T Q x^* + \bar{u}^T R u^*]\,\mathrm{d}t + \bar{x}^T(t_f) Q_f x^*(t_f) = 0 \tag{2.17a}$$

Denoting $\lambda(t)$ as the adjoint variable, which satisfies

$$\frac{\mathrm{d}\lambda}{\mathrm{d}t} = -A^T\lambda - Qx^* \qquad \lambda(t_f) = Q_f x^*(t_f) \tag{2.17b}$$

and substituting Eq. (2.17b) into Eq. (2.17a), we get

$$\int_0^{t_f} \bar{x}^T(-A^T\lambda - \dot{\lambda})\,\mathrm{d}t + \int_0^{t_f} \bar{u}^T R u^*\,\mathrm{d}t + \bar{x}^T(t_f)\lambda(t_f) = 0$$

Substituting Eq. (2.15) into the preceding equation leads to

$$\int_0^{t_f} \left(-\dot{\bar{x}}^T\lambda + \bar{u}^T B_2^T\lambda - \bar{x}^T\dot{\lambda}\right)\mathrm{d}t + \int_0^{t_f} \bar{u}^T R u^*\,\mathrm{d}t + \bar{x}^T(t_f)\lambda(t_f) = 0$$

Since

$$\frac{\mathrm{d}(x^{-T}\lambda)}{\mathrm{d}t} = \dot{\bar{x}}^T\lambda + \bar{x}^T\dot{\lambda} \qquad \bar{x}(0) = 0$$

we obtain

$$\int_0^{t_f} \bar{u}^T\left[B_2^T\lambda + Ru^*\right]\mathrm{d}t = 0$$

Because $\bar{u}$ is arbitrary, we conclude that

$$u^* = -R^{-1}B_2^T\lambda \tag{2.18a}$$

Similarly, if we take $u = u^*$, and consider the perturbation of w^* to $w = w^* + \eta\bar{w}$, we find that w^* maximizes the resulting J only if

$$w^* = \gamma^{-2}B_1^T\lambda \tag{2.18b}$$

Substituting Eqs. (2.18a) and (2.18b) into Eq. (2.10) results in the following two-point boundary-value problem (TPBVP):

$$\begin{bmatrix} \dot{x}^* \\ \dot{\lambda} \end{bmatrix} = \begin{bmatrix} A & \gamma^{-2}B_1B_1^T - B_2R^{-1}B_2^T \\ -Q & -A^T \end{bmatrix}\begin{bmatrix} x^* \\ \lambda \end{bmatrix} + \begin{bmatrix} B_3 \\ 0 \end{bmatrix} d$$

$$x^*(0) = x_0$$

$$\lambda(t_f) = Q_f x^*(t_f) \tag{2.19}$$

Hence the solution of Eqs. (2.19) leads to the necessary conditions for optimality, where the initial and terminal conditions emerged from Eqs. (2.10) and (2.17b).

Remark: The preceding results are valid for $t_0 \neq 0$ since the system is time invariant.

Example 2.1: Consider the simple scalar case

$$\dot{x} = u + w + d \qquad x(t_0) = x_0$$

with the following cost

$$\frac{1}{2}cx^2(t_f) + \int_0^{t_f} [u^2(t) - \gamma^2 w^2(t)]\, dt$$

where d is constant, $c > 0$, and $\gamma \geq 1$.

This is, in fact, the simplest possible pursuit-evasion game where the pursuer's and evader's normal velocities are directly controlled by u and w, respectively. This may correspond, for example, to missile (pursuer) and aircraft (evader) that both have very fast path-angle control loops. The fact that d is constant corresponds to the known part of the evader's normal velocity being constant. The meaning of $\gamma \geq 1$ is that the unknown part of the evasive normal velocities of the target are of smaller energy than applied by the pursuer.

The adjoint equation (2.17b) becomes

$$\dot{\lambda} = 0 \qquad \lambda(t_f) = cx^*(t_f)$$

hence

$$\lambda = cx^*(t_f)$$

From Eqs. (2.18a) and (2.18b) we get

$$u^* = -\lambda = -cx^*(t_f)$$

$$w^* = \gamma^{-2}\lambda = \gamma^{-2}cx^*(t_f)$$

Substituting the last two expressions into the state equation leads to

$$x^*(t) = x_0 + \int_0^t \left[(-1+\gamma^{-2})cx^*(t_f) + d\right] d\tau = \left[(-1+\gamma^{-2})cx^*(t_f) + d\right](t - t_0)$$

By evaluating the $x^*(t)$ at $t = t_f$ and solving for $x^*(t_f)$, we readily obtain

$$x^*(t_f) = \frac{x_0 + (t_f - t_0)d}{c\left[(1-\gamma^{-2})(t_f - t_0) + (1/c)\right]}$$

thus

$$u^*(t) = -\frac{x_0 + (t_f - t_0)d}{\left[(1-\gamma^{-2})(t_f - t_0) + (1/c)\right]}$$

Substituting $t = t_0$ we get an expression for $u^*(t_0)$ as an affine function of the present value of $x(t_0) = x_0$.

Moreover, since the problem is autonomous, we may omit the zero subscript (with the obvious interpretation of resolving the problem at different initial times). Thus we can write

$$u^*(t) = -[g_1 x(t) + g_2 d]$$

$$g_1 = \frac{1}{\left[(1-\gamma^{-2})(t_f - t) + (1/c)\right]}$$

$$g_2 = \frac{(t_f - t)}{\left[(1-\gamma^{-2})(t_f - t) + (1/c)\right]}$$

Notice that we have obtained the candidate control in feedback form. Also notice that for $c \to \infty$ we get

$$u^*(t) = -\frac{1}{(1-\gamma^{-2})(t_f - t)}x(t) - \frac{1}{1-\gamma^{-2}}d(t)$$

For $\gamma \to \infty$ (i.e., exactly known evasive actions of the target) the pursuer exactly cancels the effect of $d(t)$. For finite $\gamma \geq 1$, the pursuer chooses not to apply an exact cancellation of $d(t)$.

IV. Riccati Equation

In similar fashion to the sweep method[3] we assume that

$$\lambda(t) = P(t)x^*(t) + \theta(t) \tag{2.20}$$

Substitution of Eq. (2.20) into the TPBVP of Eq. (2.19) gives us

$$\dot{x}^* = \left[A + \left(\gamma^{-2}B_1B_1^T - B_2R^{-1}B_2^T\right)P\right]x^* + \left(\gamma^{-2}B_1B_1^T - B_2R^{-1}B_2^T\right)\theta + B_3d \tag{2.21a}$$

$$\dot{\lambda} = \dot{P}x^* + P\dot{x}^* + \dot{\theta} = -Qx^* - A^T(Px^* + \theta) \tag{2.21b}$$

Substituting Eq. (2.21a) into Eq. (2.21b), we obtain

$$\begin{aligned}&\left[\dot{P} + PA + A^TP + P\left(\gamma^{-2}B_1B_1^T - B_2R^{-1}B_2^T\right)P + Q\right]x^* \\ &\quad + \dot{\theta} + A^T\theta + PB_3d + P\left(\gamma^{-2}B_1B_1^T - B_2R^{-1}B_2^T\right)\theta = 0\end{aligned} \tag{2.22}$$

A solution that also satisfies the required end condition, that is,

$$Q_fx^*(t_f) = P(t_f)x^*(t_f) + \theta(t_f) \tag{2.23}$$

can be obtained by separating the P terms from the θ terms in Eq. (2.22), thus

$$\dot{P} + PA + A^TP + P\left(\gamma^{-2}B_1B_1^T - B_2R^{-1}B_2^T\right)P + Q = 0 \tag{2.24}$$

$$P(t_f) = Q_f \tag{2.25}$$

$$\dot{\theta} = -\left[A^T + P\left(\gamma^{-2}B_1B_1^T - B_2R^{-1}B_2^T\right)\right]\theta - PB_3d \qquad \theta(t_f) = 0 \tag{2.26}$$

To show that this solution is unique and to justify the sweep assumption, we refer the reader to Appendix A.

Summing up the preceding results, we have found explicit expressions for the candidates saddle-point control and disturbance functions:

$$u^* = -R^{-1}B_2^TPx^* - R^{-1}B_2^T\theta \tag{2.27a}$$

$$w^* = \gamma^{-2}B_1^TPx^* + \gamma^{-2}B_1^T\theta \tag{2.27b}$$

where

$$\dot{\theta} = -\left[A^T + P\left(\gamma^{-2}B_1B_1^T - B_2R^{-1}B_2^T\right)\right]\theta - PB_3d \tag{2.28a}$$

$$\theta(t_f) = 0 \tag{2.28b}$$

and where

$$-\dot{P} = A^T P + PA + P\left(\gamma^{-2} B_1 B_1^T - B_2 R^{-1} B_2^T\right)P + Q = 0 \quad (2.29a)$$

$$P(t_f) = Q_f \quad (2.29b)$$

The existence of $P(t)$ is shown in Appendix A to be equivalent to having a nonsingular $\phi_{11}(t, t_f) + \phi_{12}(t, t_f)Q_f$, where $\{\phi_{ij}(t, \tau)\}$ $i, j = 1, 2$ are partitions of the transition matrix corresponding to Eqs. (2.19) (conformally with x and λ), namely

$$\frac{d}{dt}\phi(t, \tau) = H\phi(t, \tau) \qquad \phi(\tau, \tau) = I$$

with

$$H = \begin{bmatrix} A & \gamma^{-2} B_1 B_1^T - B_2 R^{-1} B_2^T \\ -Q & -A^T \end{bmatrix}$$

hence

$$\begin{bmatrix} x(t) \\ \lambda(t) \end{bmatrix} = \begin{bmatrix} \phi_{11}(t, t_f) & \phi_{12}(t, t_f) \\ \phi_{21}(t, t_f) & \phi_{22}(t, t_f) \end{bmatrix} \begin{bmatrix} I \\ Q_f \end{bmatrix} x(t_f) \quad (2.30)$$

The existence of $P(t)$, therefore, should be investigated to complete the necessity proof. For this purpose we introduce the well-known concept of a conjugate point.

Definition: Two times t_0 and t_f with $t_0 \leq t_f$ are conjugate points of the TPBVP [Eqs. (2.19)] if there is a nontrivial solution to Eqs. (2.19) with $d = 0$ such that $x(t_0) = 0$ and $\lambda(t_f) = Q_f x(t_f)$.

Claim: A conjugate point does not exist if and only if $\phi_{11}(t, t_f) + \phi_{12}(t, t_f)Q_f$ is nonsingular.

Proof: The proof is given in Ref. 2, but we also give it here for the sake of completeness. It readily follows from the definition of conjugate points that by substituting $x(t_0) = 0$ and $\lambda(t_f) = Q_f x(t_f)$ we get $0 = [\phi_{11}(t, t_f) + \phi_{12}(t, t_f)Q_f]x(t_f)$, which implies that $\phi_{11}(t, t_f) + \phi_{12}(t, t_f)Q_f$ is singular. Conversely, assume that there exists a nonzero g such that at some $t_0 \leq t_f$ we have $0 = [\phi_{11}(t_0, t_f) + \phi_{12}(t_0, t_f)Q_f]g$. Examining the case when $x(t_f) = g$, we get from Eq. (2.30) at $t = t_0$ that $x(t_0) = 0$—thus a conjugate point is obtained. QED

Therefore, we have shown the equivalence of the nonexistence of a conjugate point and the nonsingularity of $\phi_{11}(t, t_f) + \phi_{12}(t, t_f)Q_f$ [which in turn is equivalent to the existence of $P(t)$]. At this point we would like to conjecture that if there exists a saddle point, then there cannot exist a conjugate point. It turns out that this conjecture is not true and, as shown by Ref. 1, a saddle point can survive a conjugate point. Thus the existence of $P(t)$ is not a necessary condition for the existence of saddle-point solutions. In Sec. V we will show that it is a sufficient condition for optimality.

Example 2.2: Consider again the preceding first-order system

$$\dot{x} = u + w + d \qquad x(t_0) = x_0$$

with the following cost

$$\frac{1}{2}cx^2(t_f) + \int_{t_0}^{t_f} [u^2(t) - \gamma^2 w^2(t)]\,\mathrm{d}t$$

The associated Riccati equations (2.28–2.29) are here

$$\dot{p} = p^2(1-\gamma^{-2}) \qquad p(t_f) = c$$

$$\dot{\theta} = -p(1-\gamma^{-2})\theta - pd \qquad \theta(t_f) = 0$$

The solution for p can be easily obtained by using

$$\frac{\mathrm{d}p}{p^2} = (1-\gamma^{-2})\,\mathrm{d}t$$

thus

$$p = \frac{1}{(1/c) + (1-\gamma^{-2})(t_f - t)}$$

We can solve now the (linear) differential equation for θ to obtain

$$\theta = \frac{(t_f - t)d}{(1/c) + (1-\gamma^{-2})(t_f - t)}$$

The candidates for the optimal control and the optimal disturbance are [Eqs. (2.27a) and (2.27b)]

$$u^* = -px^* + \theta$$

$$w^* = \gamma^{-2}(px^* - \theta)$$

hence again

$$u^* = -\frac{x^* + (t_f - t)d}{\left[(1-\gamma^{-2})(t_f - t) + (1/c)\right]} \qquad w^* = -\gamma^{-2}u^*$$

For $c > 0$ and $\gamma \geq 1$ there is no finite escape time of p and the control is bounded with no conjugate points.

V. Sufficient Condition

Assume that there exists a P that satisfies the Riccati equation (2.24) for $t \in [0, t_f]$ with the terminal condition Eq. (2.25), and consider an M such that

$$M \equiv x_0^T P(0)x_0 - x^T(t_f)Q_f x(t_f) + \int_0^{t_f} \frac{d}{\mathrm{d}t}(x^T P x)\,\mathrm{d}t \tag{2.31}$$

Note that $P(t_f) = Q_f$ yields $M = 0$.

Using the chain rule for the time derivative we have the following from Eq. (2.31):

$$\frac{d}{\mathrm{d}t}(x^T P x) = (\dot{x}^T P x) + (x^T \dot{P} x) + (x^T P \dot{x}) \tag{2.32}$$

Also, we arrive at

$$M = x_0^T P(0)x_0 - x^T(t_f)Q_f x(t_f) - \int_0^{t_f} (x^T Qx + u^T Ru - \gamma^2 w^T w)\,\mathrm{d}t$$
$$+ \int_0^{t_f} \left(u + R^{-1}B_2^T Px\right)^T R\left(u + R^{-1}B_2^T Px\right)\mathrm{d}t$$
$$- \gamma^2 \int_0^{t_f} \left(w - \gamma^{-2}B_1^T Px\right)^T \left(w - \gamma^{-2}B_1^T Px\right)\mathrm{d}t + 2\int_0^{t_f} \left(x^T PB_3 d\right)\mathrm{d}t \tag{2.33}$$

Identifying the second and the third terms in the last expression as $-J$ [see Eq. (2.3)], we obtain

$$J(u, w) = x_0^T P(0)x_0 + \int_0^{t_f} \left(u + R^{-1}B_2^T Px\right)^T R\left(u + R^{-1}B_2^T Px\right)\mathrm{d}t$$
$$- \gamma^2 \int_0^{t_f} \left(w - \gamma^{-2}R^{-1}B_1^T Px\right)^T \left(w - \gamma^{-2}R^{-1}B_1^T Px\right)\mathrm{d}t$$
$$+ 2\int_0^{t_f} \left(x^T PB_3 d\right)\mathrm{d}t \tag{2.34}$$

For $d = 0$, we have[2]

$$J(u, w) = x_0^T P(0)x_0 + \int_0^{t_f} (u - u^*)^T R(u - u^*)\,\mathrm{d}t - \gamma^2 \int_0^{t_f} (w - w^*)^T (w - w^*)\,\mathrm{d}t \tag{2.35}$$

Sufficiency is obvious since any deviation from the optimal strategy is penalized; therefore, $J(u^*, w) \le J(u^*, w^*) \le J(u, w^*)$ where the value of the game is

$$J(u^*, w^*) = x_0^T P(0)x_0 \tag{2.36}$$

For nonzero d the proof is somewhat more intricate and we need to develop the last term in Eq. (2.33). Let Θ be a solution to Eq. (2.26), then

$$x^T PB_3 d = -x^T\dot{\Theta} - x^T P\left(\gamma^{-2}B_1 B_1^T - B_2 R^{-1}B_2^T\right)\Theta - x^T A^T \Theta$$
$$= -\frac{d}{\mathrm{d}t}(x^T\Theta) + \dot{x}^T\Theta - x^T P\left(\gamma^{-2}B_1 B_1^T - B_2 R^{-1}B_2^T\right)\Theta - x^T A^T\Theta \tag{2.37}$$

Substituting the state equation (2.1) for the $\dot{x}$ term, and after some algebraic manipulation, we notice that

$$x^T PB_3 d = -\frac{d}{\mathrm{d}t}(x^T\Theta) + (w - w^*)^T B_1^T\Theta + (u - u^*)^T B_2^T\Theta + d^T B_3^T\Theta$$
$$+ \Theta^T\left(\gamma^{-2}B_1 B_1^T - B_2 R^{-1}B_2^T\right)\Theta \tag{2.38}$$

Using the zero terminal conditions for Θ, we get

$$-2\int_0^{t_f} \frac{d}{dt}(x^T\Theta)\,dt = 2x_0^T\Theta(0) \tag{2.39}$$

Denoting the following expression for J_0

$$\begin{aligned} J_0 = x_0^T[P(0)x_0 + 2\Theta(0)] + 2\int_0^{t_f} d^T B_3^T \Theta\,dt \\ + \int_0^{t_f} \theta^T(\gamma^{-2}B_1B_1^T - B_2R^{-1}B_2^T)\theta\,dt \end{aligned} \tag{2.40}$$

we obtain

$$\begin{aligned} J(u,w) = J_0 + \int_0^{t_f} \left(u - u^* - R^{-1}B_2^T\Theta\right)^T R\left(u - u^* - R^{-1}B_2^T\Theta\right) dt \\ -\gamma^2\int_0^{t_f} \left(w - w^* + \gamma^{-2}B_1^T\Theta\right)^T\left(w - w^* + \gamma^{-2}B_1^T\Theta\right) dt \\ +2\int_0^{t_f} \left[(w - w^*)^T B_1^T\Theta + (u - u^*)^T B_2^T\Theta\right] dt \\ + \int_0^{t_f} \Theta^T\left(\gamma^{-2}B_1B_1^T - B_2R^{-1}B_2^T\right)\Theta\,dt \end{aligned} \tag{2.41}$$

or simply

$$J(u,w) = J_0 + \int_0^{t_f} (u - u^*)^T R(u - u^*)\,dt - \gamma^2\int_0^{t_f} (w - w^*)^T(w - w^*)\,dt \tag{2.42}$$

Thus $J(u^*, w^*) = J_0$ and sufficiency is proven.

VI. Adjoint Analysis

The following analysis is an efficient and effective way of evaluating control strategies. The origin of this method goes back to Volterra circa 1870 but it was popularized only in the 1950s by Laninig and Batin.[4] In Ref. 4, the method is presented by reversing the block diagram of the Laplace transform representations. We will, however, follow the state-space development of Kelley,[5] which is more suitable for our purposes and fits nicely into the general framework of the present monograph.

Assuming Eq. (2.1) with optimal strategies (2.27) being played by both sides, we get

$$\dot{x} = \left[A + \gamma^{-2}B_1B_1^TP - B_2R^{-1}B_2^TP\right]x + \gamma^{-2}B_1B_1^T\Theta - B_2R^{-1}B_2^T\Theta + B_3d \tag{2.43}$$

Notice that in general d, P, and Θ are all time varying and can be computed from t_f backward for all $t \in [0, t_f]$.

Define $\bar{A}(t)$ and $\bar{d}(t)$ by

$$\bar{A}(t) \equiv \left[A + \gamma^{-2}B_1B_1^TP - B_2R^{-1}B_2^TP\right] \tag{2.44a}$$

$$\bar{d}(t) \equiv \gamma^{-2}B_1B_1^T\Theta - B_2R^{-1}B_2^T\Theta + B_3d \tag{2.44b}$$

Remark: If the target is not playing optimally, so that the unmeasured evasive maneuver w is not in feedback form, then we will use

$$\bar{A}(t) \equiv \left[A - B_2 R^{-1} B_2^T P\right] \tag{2.44c}$$

$$\bar{d}(t) \equiv -B_2 R^{-1} B_2^T Q + B_1 w + B_3 d \tag{2.44d}$$

In any case the state equation gets the form

$$\dot{x} = \bar{A}(t)x + \bar{d}(t) \qquad x(0) = x_0 \tag{2.45}$$

In the adjoint analysis we consider the following adjoint system

$$\dot{\lambda} = -\bar{A}^T(t)\lambda \tag{2.46}$$

with the terminal conditions

$$\lambda(t_f) = [1, 0, \ldots, 0]^T \tag{2.47}$$

It readily follows that

$$\frac{d}{dt}(\lambda^T x) = \dot{\lambda}^T x + \lambda^T \dot{x} = -[\bar{A}^T(t)\lambda]^T x + \lambda^T[\bar{A}(t)x + \bar{d}(t)] = \lambda^T \bar{d}(t) \tag{2.48}$$

and thus we have

$$\int_0^{t_f} \frac{d}{dt}(\lambda^T x)\,dt = \int_0^{t_f} \lambda^T \bar{d}(t)\,dt \tag{2.49}$$

Because of the special structure of $\lambda(t_f)$, we find that

$$x_1(t_f) = \lambda^T(0)x_0 + \int_0^{t_f} \lambda^T(t)\bar{d}(t)\,dt \tag{2.50}$$

where x_1 denotes the first entry of x (in most cases, the relative separation, which at the final time is the miss distance—see Chapter 1).

We have obtained an expression for the final value of x_1 without calculating the entire state trajectory. (A similar calculation with $\lambda^T(t_f) = [0, 1, 0, \ldots, 0]$ would yield the value of the second state variable at the terminal time, etc.) However, we still need to calculate the adjoint function $\lambda(t)$. In general, one prefers dealing with initial rather than terminal value problems. To this end, we consider Eqs. (2.46) and (2.47) in reversed time using $\tau \equiv t_f - t$. Defining $\bar{\lambda}(\tau)$ as

$$\bar{\lambda}(\tau) = \lambda(t_f - \tau) \tag{2.51}$$

we get from Eq. (2.46) that

$$\frac{d\bar{\lambda}(\tau)}{d\tau} = \bar{A}^T(t_f - \tau)\bar{\lambda}(\tau) \tag{2.52}$$

with the initial conditions

$$\bar{\lambda}(0) = [1, 0, \ldots, 0]^T \tag{2.53}$$

If in Eq. (2.50) we make the required change in t, that is, $t = t_f - \tau$, and use Eq. (2.51) we have

$$x_1(t_f) = \bar{\lambda}^T(t_f)x_0 + \int_0^{t_f} \bar{\lambda}^T(\tau)\bar{d}(t_f - \tau)\,\mathrm{d}\tau \tag{2.54}$$

The last expression may be viewed as a Green's (or influence) function for the adjoint variable.

More importantly, by solving the adjoint equations (2.52–2.54) (either analytically or by numerical simulation), one obtains a family of solutions to the original problem for different t_f. This implies a tremendous saving in simulation runs. Suppose, for example, that $\bar{d} = 0$ (no target maneuvers). For this simple case, the scalar product of the vector function $\bar{\lambda}(t_f)$ with the initial condition vector x_0 provides the required terminal value. Hence, once we obtain the influence function $\bar{\lambda}(\cdot)$ over a domain of interest (e.g., by a single simulation run), we have the endgame results for all of the t_f in this domain.

It is even more impressive (and indeed useful) to consider the following special case. Let $d = 0$ (no measurements of target's acceleration are available to the pursuer) and let w play nonoptimally, by initiating a step function (i.e., constant acceleration) at time $t_f - t_{\mathrm{go}}$ to evade the missile. Thus, Eqs. (2.44a) and (2.44b) become

$$\bar{A}(t) \equiv \left[A - B_2 R^{-1} B_2^T P\right] \tag{2.55a}$$

$$w(t) \equiv \begin{cases} 0 & t < t_f - t_{\mathrm{go}} \\ 1 & t \geq t_f - t_{\mathrm{go}} \end{cases} \tag{2.55b}$$

Using $\bar{d}(t) = B_1 w(t)$, we can write

$$\bar{d}(t_f - t) = \begin{cases} 0 & t_{\mathrm{go}} < t \\ B_1 & t_{\mathrm{go}} \geq t \end{cases} \tag{2.56}$$

Hence if $x_0 = 0$ as well, we get a simple expression for $x_1(t_f)$, that is,

$$x_1(t_f) = \int_0^{t_f} \bar{\lambda}^T(t)\bar{d}(t_f - t)\,\mathrm{d}t = \int_0^{t_{\mathrm{go}}} \bar{\lambda}^T(t)B_1\,\mathrm{d}t \tag{2.57}$$

which relates the terminal value of interest (in most cases the miss distance) to t_{go}, the time-to-go remaining for interception after the target has initiated an evasive maneuver.

Remark: Another interesting case is the one where $x_0 = 0$, $d = 0$, and w is a zero-mean white noise with intensity R for $t \geq t_f - t_{\mathrm{go}}$ and zero elsewhere. In such a case we have $E[w(t_1)w(t_2)] = R\delta(t_1 - t_2)$ for $t_f \geq t_2 \geq t_1 \geq t_f - t_{\mathrm{go}}$ where E is the ensemble mean over the statistics of w and δ is Dirac's function. It follows from Eq. (2.54) that

$$x_1(t_f) = \int_0^{t_f} \bar{\lambda}^T(\tau)\bar{d}(t_f - \tau)\,\mathrm{d}\tau \tag{2.58}$$

where

$$\bar{d}(t) = \begin{cases} B_1 w(t) & t \geq t_f - t_{\mathrm{go}} \\ 0 & \text{elsewhere} \end{cases}$$

Thus

$$\bar{d}(t-\tau)=\begin{cases} B_1 w(t_f-\tau) & \tau \le t_{\mathrm{go}} \\ 0 & \text{elsewhere} \end{cases}$$

and, therefore, can write

$$x_1(t_f)=\int_0^{t_{\mathrm{go}}} \bar{\lambda}^T(\tau) w(t_f-\tau)\,\mathrm{d}\tau \tag{2.59}$$

By taking the mean square of both sides, we obtain

$$E\left[x_1^2(t_f)\right]=E\left(\int_0^{t_{\mathrm{go}}}\int_0^{t_{\mathrm{go}}} \bar{\lambda}^T(\tau)B_1 w(t_f-\tau)w(t_f-\mu)B_1^T\bar{\lambda}(\tau)\,\mathrm{d}\mu\,\mathrm{d}\tau\right)$$

We may change the order of integration and expectation (to justify this nontrivial fact, see Theorem 4.1 in Ref. 6) to get

$$\begin{aligned} E\left[x_1^2(t_f)\right] &= \int_0^{t_{\mathrm{go}}}\int_0^{t_{\mathrm{go}}} \bar{\lambda}^T(\tau)B_1 E[w(t_f-\tau)w(t_f-\mu)]B_1^T\,\bar{\lambda}(\mu)\,\mathrm{d}\mu\,\mathrm{d}\tau \\ &= \int_0^{t_{\mathrm{go}}}\int_0^{t_{\mathrm{go}}} \bar{\lambda}^T(\tau)B_1 R\delta(\mu-\tau)B_1^T\,\bar{\lambda}(\mu)\,\mathrm{d}\mu\,\mathrm{d}\tau \\ &= \int_0^{t_{\mathrm{go}}} \bar{\lambda}^T(\tau)B_1 R B_1^T\,\bar{\lambda}(\tau)\,\mathrm{d}\tau = R\int_0^{t_{\mathrm{go}}}\left[\bar{\lambda}^T(\tau)B_1\right]^2\mathrm{d}\tau \end{aligned} \tag{2.60}$$

Example 2.3: Consider a restricted version of the preceding scalar case, where $\gamma \to \infty$, that is, the control is designed with the assumption of $w = 0$ in the following dynamic equation

$$\dot{x}=u+w+d \qquad x(0)=0$$

with the cost function

$$J=\frac{c}{2}x^2(t_f)+\int_0^{t_f} u^2(t)\,\mathrm{d}t$$

If, after the design process, it turns out that $w \neq 0$, then we must distinguish between the measured target maneuver d and the unmeasured one w. We will consider two cases: case 1 has a unit constant unmeasured evasive maneuver ($w=1$ and $d=0$), and case 2 has a unit constant measured evasive maneuver ($w=0$ and $d=1$).

In both cases $\bar{A}(t)$ satisfies

$$\bar{A}(t)=-p=\frac{-1}{(1/c)+(t_f-t)}$$

and, therefore, the adjoint equation becomes

$$\frac{\mathrm{d}\bar{\lambda}(\tau)}{\mathrm{d}\tau}=-\frac{\bar{\lambda}(\tau)}{(1/c)+\tau} \qquad \bar{\lambda}(0)=1$$

The solution can be obtained easily as

$$\bar{\lambda}(\tau)=\frac{1}{1+\tau c}$$

In case 1 (unmeasured target's maneuver) we have

$$\bar{d}(t) = w(t) = 1$$

and thus from Eq. (2.57) we get

$$x(t_f) = \int_0^{t_f} \bar{\lambda}(t) B_1 \, \mathrm{d}t = \frac{1}{c} \ln(1 + ct_f)$$

Notice that in this case the closed-loop state equation becomes

$$\dot{x} = -\frac{x}{(1/c) + (t_f - t)} + 1$$

hence

$$x(t) = [(1/c) + (t_f - t)] \ln\left[\frac{1 + ct_f}{1 + c(t_f - t)}\right]$$

and the final value $x(t_f)$ is obtained as predicted by the adjoint analysis.

In case 2 we have

$$\bar{d}(t) = -\theta(t) + d(t) = -\frac{(t_f - t)}{(1/c) + (t_f - t)} + 1$$

hence

$$\bar{d}(t_f - \tau) = \frac{1}{1 + \tau c}$$

and therefore

$$x(t_f) = \int_0^{t_f} \bar{\lambda}(\tau) \bar{d}(t_f - \tau) \, \mathrm{d}\tau = \int_0^{t_f} \left(\frac{1}{1 + \tau c}\right)^2 \mathrm{d}\tau = \frac{t_f}{1 + ct_f}$$

Again we may integrate directly the resulting state equation

$$\dot{x} = -\frac{x}{(1/c) + (t_f - t)} - \frac{t_f - t}{(1/c) + (t_f - t)} + 1$$

to obtain

$$x(t) = \frac{t}{1 + ct_f}$$

with the final value $x(t_f)$ as predicted by the adjoint analysis.

Figure 2.1 compares the miss distances of the two cases for various values of conflict duration where we let $c = 100$. The results clearly demonstrate the advantage obtained by measuring and feeding back the target acceleration.

Notice that if one is interested in final values only, the adjoint analysis provides the complete answers for various values of conflict duration (i.e., time-to-go from the evasive-maneuver initiation) without having to determine the entire trajectories.

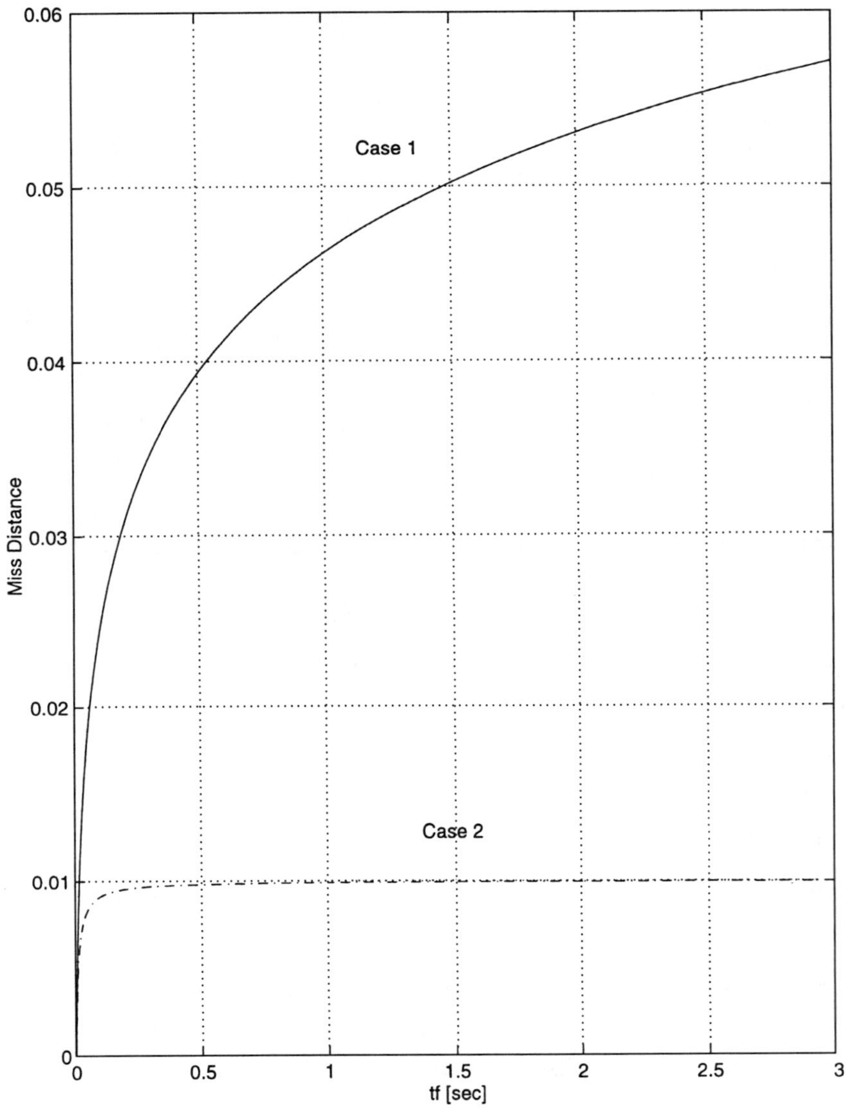

Fig. 2.1 Miss distance with and without target acceleration feedback.

VII. Summary and Conclusions

An LQ differential game problem was formulated and analyzed. The guidance problems of this book will be presented as special cases of this problem or as various extensions to it. The TPBVP, in the form of state–costate equations, is shown to be a necessary condition for optimality. The Riccati equation, on the other hand, provides a sufficient condition for optimality. Searching for optimal candidates can be, and in fact will be, performed by both representations—as was demonstrated by a representative example. Finally the adjoint analysis, which will be used frequently in the following chapters for performance evaluation, was presented and demonstrated.

References

[1]Basar, T., and Bernhard, P., *H^∞-Optimal Control and Related Minimax Design Problems*, 2nd ed., Birkhäuser, Boston, 1991, Chap. 4.

[2]Green, M., and Limebeer, D. J. N., *Linear Robust Control*, Prentice-Hall, Englewood Cliffs, NJ, 1995, Chap. 6.

[3]Bryson, A. E., and Ho, Y. C., *Applied Optimal Control*, Hemisphere, New York, 1975, Chap. 5.

[4]Zarchan, P., *Tactical and Strategic Missile Guidance*, Vol. 124, Progress in Astronautics and Aeronautics, AIAA, Washington, DC, 1990, Chap. 3.

[5]Kelley, H. J., "Methods of Gradients," *Optimization Techniques*, edited by G. Leitmann, Vol. 5, Mathematics in Science and Engineering, Academic Press, New York, 1962, pp. 218–220.

[6]Jazwinski, A. H., *Stochastic Processes and Filtering Theory*, Vol. 64, Mathematics in Science and Engineering, Academic Press, New York, 1970, Chap. 4.

Chapter 3

Optimal Guidance Based on Linear-Quadratic One-Sided Optimization

I. Introduction

MODERN guidance techniques are mainly based on optimal control theory and on differential games. After comparing these two approaches, the use of differential game guidance methods was found to be preferable—particularly against maneuvering targets.[1] However, PN, which is in widespread use, as well as other so-called "advanced guidance methods,"[2] are optimal strategies, obtained from LQ one-sided optimization problems where the cost is the missile's control effort and a zero miss is imposed as a boundary condition. Traditionally, these results have been obtained through the use of Schwartz's inequality.[2] We will derive these well-known results by the state-space formulation of the previous chapter.

We will use the linear model of planar pursuit-evasion conflicts as presented in Chapter 1. Several problems will be formulated and analytically solved in increasing order of complexity. First, an ideal pursuer chasing a nonmaneuverable target will be considered, with and without terminal velocity constraints. Using this result, we will rederive the fact that PN is an optimal strategy. Next, the target will be allowed to maneuver with constant acceleration, and the assumption of an ideal time response will be removed and replaced by a first-order time lag model for the pursuer. Finally, for the sake of completeness, a first-order time lag will also be assumed for the evader, and the problem will be solved analytically. Numerical comparisons will be presented for the various methods based on forward and adjoint simulations.

As previously noted, different derivations for most of the described cases do exist in the literature. However, the ensuing derivation as well as the resulting optimal strategies are more general in nature so that new results—some of which are presented in the next few chapters—can be obtained along the same lines.

II. Problem Formulation

The following state-space formulation was obtained in Chapter 1:

$$\dot{x}_1 = x_2 \qquad \dot{x}_2 = u + w \tag{3.1}$$

or in matrix form

$$\dot{x} = Ax + Bu + Dw \tag{3.2}$$

where

$$A = \begin{bmatrix} 0 & 1 \\ 0 & 0 \end{bmatrix} \qquad B = \begin{bmatrix} 0 \\ 1 \end{bmatrix} \qquad D = \begin{bmatrix} 0 \\ 1 \end{bmatrix}$$

Recall that x_1 and x_2, the components of x, are the relative displacement and velocity, respectively, and u and w are the corresponding normal accelerations of the pursuer and the evader.

The optimal control problem is the following:

$$\min_u J = \frac{b}{2}x_1^2(t_f) + \frac{c}{2}x_2^2(t_f) + \frac{1}{2}\int_{t_0}^{t_f} u^2(t)\,\mathrm{d}t \tag{3.3}$$

where $b > 0$ is the penalty imposed by the miss distance (i.e., the relative separation at the final time). Typically we let $c = 0$; however, we may also consider cases where the pursuer's objective is to minimize the terminal lateral velocity ($c > 0$), or even cases where the pursuer's objective is to maximize this velocity ($c < 0$). This objective depends heavily on the particular conflict for which the problem is formulated.

Naturally, for perfect intercepts $[x_1(t_f) = 0]$, we will require that

$$b \to \infty \tag{3.4}$$

and for perfect rendezvous,[3] where $x_1(t_f) = 0$ and $x_2(t_f) = 0$, we will require that

$$b \to \infty \qquad c \to \infty \tag{3.5}$$

III. Nonmaneuvering Target

A. General Solution

In this simple case the state equation is

$$\dot{x} = Ax + Bu \qquad x^T(t_0) = [x_1(t_0), x_2(t_0)]$$

$$A = \begin{bmatrix} 0 & 1 \\ 0 & 0 \end{bmatrix} \qquad B = \begin{bmatrix} 0 \\ 1 \end{bmatrix} \tag{3.6}$$

where $x(t_0)$ is a fixed unknown target vector. Note that $[x_1(t_0) + x_2(t_0)(t_f - t_0)]/R$ is the initial heading error.

First-order necessary conditions for optimality are the following adjoint equations [Eqs. (2.17b)]:

$$\dot{\lambda}_1 = 0 \qquad \dot{\lambda}_2 = -\lambda_1 \qquad u = -\lambda_2 \tag{3.7}$$

The terminal conditions are

$$\lambda_1(t_f) = bx_1(t_f) \qquad \lambda_2(t_f) = cx_2(t_f)$$

Thus

$$\lambda_1 = bx_1(t_f) \qquad \lambda_2 = bx_1(t_f)(t_f - t) + cx_2(t_f) \tag{3.8}$$

Substituting $u(t) = -[bx_1(t_f)(t_f - t) + cx_2(t_f)]$ into Eqs. (3.6) and integrating it from t_0 to t, we get the following expression for $x_1(t)$ and $x_2(t)$:

$$x_1(t) = \left[\frac{1}{2}b\left(\frac{t^3}{3} - \frac{t_0^3}{3}\right) - \frac{1}{2}bt_0^2(t - t_0)\right]x_1(t_f) - \frac{1}{2}b(t - t_0)^2 t_f x_1(t_f)$$

$$-\frac{1}{2}c(t - t_0)^2 x_2(t_f) + x_1(t_0) + x_2(t_0)(t - t_0) \tag{3.9}$$

$$x_2(t) = \tfrac{1}{2}b\left(t^2 - t_0^2\right)x_1(t_f) - b(t - t_0)x_1(t_f) - c(t - t_0)x_2(t_f) + x_2(t_0)$$

Evaluating Eqs. (3.9) at $t = t_f$ we can solve the resulting linear algebraic equations for the unknown terminal values $x_1(t_f)$ and $x_2(t_f)$ and thus arrive at an expression for $u(t)$ and in particular an expression for $u(t_0)$, which is linear in $x_1(t_0)$ and $x_2(t_0)$, that is

$$u(t_0) = -[g_1 x_1(t_0) + g_2 x_2(t_0)]$$

where

$$g_1 = \frac{b\tau\left(\frac{1}{2}\tau c + 1\right)}{\tau^4 \frac{cb}{12} + \tau c + \tau^3 \frac{b}{3} + 1}$$

$$g_2 = \frac{\frac{1}{3}\tau^3 cb + \tau^2 b + c}{\tau^4 \frac{cb}{12} + \tau c + \tau^3 \frac{b}{3} + 1} \tag{3.10a}$$

and where $\tau = t_f - t_0$.

Since t_0 is arbitrary, we may replace t_0 by t and thus obtain the following feedback form

$$u(t) = -[g_1 x_1(t) + g_2 x_2(t)]$$

where

$$g_1 = \frac{b\tau\left(\frac{1}{2}\tau c + 1\right)}{\tau^4 \frac{cb}{12} + \tau c + \tau^3 \frac{b}{3} + 1}$$

$$g_2 = \frac{\frac{1}{3}\tau^3 cb + \tau^2 b + c}{\tau^4 \frac{cb}{12} + \tau c + \tau^3 \frac{b}{3} + 1} \tag{3.10b}$$

and where $\tau = t_f - t$.

Listing 3.1 is a MAPLE routine that performs the task of deriving Eqs. (3.7)–(3.10) automatically. (For more information about MAPLE see Appendix B.)

For this simple case the derivation can be obtained by hand calculation. However, the more complicated cases render the employment of such a symbolic code indispensable.

If we formulate the problem in the following way

$$J = \frac{1}{2}x^T(t_f)P(t_f)x(t_f) + \frac{1}{2}\int_{t_0}^{t_f} u^T(t)u(t)\,\mathrm{d}t$$

Listing 3.1 Gains for ideal pursuer and stationary target

```
> #listing 3.1
> # l1(t) is constant and equals b*x1f.
> # the following is the remaining costate equation for l2(t), namely d(l2)/dt =−l1.
> #
> sys:=diff(l2(t), t)+b*x1f;
> dsolve({sys,l2(tf)=c*x2f},l2(t));
> assign(");
> #
> # the following is the state equation for x2(t), namely d(x2)/dt=u=−l2.
> #
> sys1:=diff(x2(t), t)+l2(t);
> dsolve({sys1,x2(t0)=v0},x2(t));
> assign(");
> #
> # the following is the state equation of x1(t), namely d(x1)/dt=x2.
> #
> sys3:=diff(x1(t), t)−x2(t);
> dsolve({sys3, x1(t0)=x0}, x1(t));
> assign(");
> #
> # the terminal conditions for x1(t) and x2(t) are as follows
> #
> x1(t);
> y1:=subs(t=tf,");
> x2(t);
> y2:=subs(t=tf,");
> #
> # solving for explicit expressions for x1f and x2f.
> #
> sys5:={x1f−y1, x2f−y2};
> solve(",{x1f,x2f});
> assign(");
> #
> # evaluating the control u=−l2 at t=t0
> #
> u:=−eval(l2(t));
> u0:=sus(t=t0,");
> simplify(");
> #
> # limit value for no velocity constraints
> #
> limit(", c=0);
> #
> # limit value for perfect intercept
> #
> limit(", b=infinity);
```

then we notice that the problem can be solved by employing the associated Riccati equation (see Chapter 2):

$$-\dot{P} = PA + A^T P - PBB^T P \tag{3.11a}$$

with the terminal conditions

$$P(t_f) = \begin{bmatrix} b & 0 \\ 0 & c \end{bmatrix}$$

The optimal strategy then takes the form

$$u = -B^T P x = -(g_1 x_1 + g_2 x_2)$$

Since $B^T = [0 \ 1]$, we identify $\{g_1, g_2\}$ with the entries of the last row of P. Thus if (and only if) these gains are bounded for all $0 < \tau \leq t_f$, there exists no conjugate point on this interval, and the solution, therefore, satisfies sufficient conditions for optimality (see Chapter 2). For $c \geq 0$, the gains are clearly bounded and therefore optimal. Notice, however, that this fact is not guaranteed for $c < 0$ since $P(t_f)$ is no longer positive definite (see also Ref. 4). To facilitate the solution of the Riccati matrix we may solve the associated Lyapunov equation for $S \equiv P^{-1}$ that satisfies

$$\dot{S} = AS + SA^T - BB^T \tag{3.11b}$$

(To prove the preceding, notice that $SP = I$, hence $\dot{S}P + S\dot{P} = 0$ and thus we obtain the corresponding Lyapunov equation.)

Listing 3.2 is a MAPLE routine that solves this equation and produces an analytical expression for the control.

B. Case Without Velocity Weighting: PN

In this case $c = 0$ and from Eqs. (3.10b) we get

$$g_1 = \frac{\tau}{(\tau^3/3) + (1/b)} \qquad g_2 = \frac{\tau^2}{(\tau^3/3) + (1/b)} \tag{3.12a}$$

Taking $b \to \infty$, we get

$$g_1 = 3/\tau^2 \qquad g_2 = 3/\tau \tag{3.12b}$$

Using the definition of x (i.e., $x_1 = y$ and $x_2 = \dot{y}$) we obtain

$$u = -\frac{N'(y + \dot{y}\tau)}{\tau^2} \qquad N' = 3$$

We have previously shown that the LOS rate is given by

$$\dot{\lambda} = \frac{y + \dot{y}\tau}{V_c \tau^2}$$

where the closing velocity is

$$V_c = V_p \cos(\gamma_{p_0}) - V_e \cos(\gamma_{e_0}) \tag{3.13}$$

Listing 3.2 Solving the Lyapunov equation—ideal case

```
> #listing 3.2
> with(linalg);
> #
> # state transition and control matrices
> #
> A:=array([[0,1],[0,0]]);
> B:=array(1..2,1..1,[[0],[1]]);
> #
> # constructing the Lyapunov equation for S=P^(−1), namely:
> # dS/dt = AS+S trans(A) − B trans(B)
> #
> S(t):=array(1..2,1..2,[[S1(t),S2(t)],[S2(t),S3(t)]]);
> temp:=evalm((S(t)&*transpose(A))+(A)&*S(t));
> SD:=evalm((transpose(temp)−(B&*transpose(B))));
> convert(SD,array);
> #
> # solving the Lyapunov equation
> #
> system:=diff(S1(t),t)=SD[1,1],diff(S2(t),t)=SD[1,2],diff(S3(t),t)=SD[2,2];
> system1:=S3(Tf)=1/c,S2(Tf)=0,S1(Tf)=1/b;
> dsolve({system,system1}, {S1(t),S2(t),S3(t)});
> assign(");
> simplify(");
> #
> # inverting S to find P and calculating the feedback gain trans(B) P
> #
> S(t):=array(1..2,1..2,[[S1(t),S2(t)],[S2(t),S3(t)]]);
> P(t):=inverse(S(t));
> G:=evalm(transpose(B)&*P(t));
> G1:=subs(c=0,");
```

Using the last two relations we find that

$$u = -N'V_c\dot{\lambda} \qquad N' = 3 \tag{3.14}$$

which is the well-known PN law.

In this case the corresponding equations of motion become

$$\begin{aligned} \dot{x}_1 &= x_2 \\ \dot{x}_2 &= -(g_1x_1 + g_2x_2) = \frac{-3}{\tau^2}x_1 + \frac{-3}{\tau}x_2 \end{aligned} \tag{3.15}$$

and we guess at a solution of the form[5] [recall that $\tau = t_f - t$, hence $\mathrm{d}()/\mathrm{d}t = -\mathrm{d}()/\mathrm{d}\tau$]

$$x_1 = c_1\tau^{r_1} + c_2\tau^{r_2} \qquad x_2 = -c_1r_1\tau^{r_1-1} - c_2r_2\tau^{r_2-1}$$

which we substitute into Eqs. (3.15) to obtain $r_1 = 1$ and $r_2 = 2$. For this simple case, there exists a closed-form expression for the trajectories

$$x_1 = c_1\tau + c_2\tau^2 \qquad x_2 = -c_1 - 2c_2\tau \tag{3.16}$$

The constants c_1 and c_2 can be determined from the initial conditions as follows:

$$\begin{aligned} c_1 &= \frac{2}{t'_f}x_1(t_0) + x_2(t_0) \\ c_2 &= \frac{-1}{t'^2_f}x_1(t_0) - \frac{1}{t'_f}x_2(t_0) \end{aligned} \tag{3.17}$$

where $t'_f = t_f - t_0$.

Note that $x_1(t_f) = 0$ (i.e., for nonmaneuvering targets, a zero miss is obtained), but $x_2(t_f) = -c_1$. If $x_1(t_0) = 0$, then $x_2(t_f) = -x_2(t_0)$, meaning that the hit angle is equal (in magnitude) to the initial heading error. If we want to lessen the hit angle, we should penalize the terminal velocity in the cost function.

C. Perfect Intercept with Terminal Velocity Weighting: Optimal Rendezvous (OR)

Assume now that $c \neq 0$. Letting $b \to \infty$ in Eqs. (3.10) [i.e., imposing $x_1(t_f) = 0$] and using L'Hôpital's rule, we obtain

$$\begin{aligned} g_1 &= \frac{6\tau + (12/c)}{\tau^3 + (4\tau^2/c)} \\ g_2 &= \frac{4\tau + (12/c)}{\tau^2 + (4\tau/c)} \end{aligned} \tag{3.18}$$

For the case $c > 0$, the solution is bounded for all $\tau \in (0, t_f]$. For $c < 0$ the solution is bounded only if

$$\tau < -\frac{4}{c}$$

Since the maximum value on the left-hand side is t_f, we obtain the condition

$$t_f < -\frac{4}{c} \tag{3.19}$$

From Sec. III.A, this inequality is the condition for no conjugate point.

The perfect rendezvous is obtained from Eqs. (3.10) by letting $b \to \infty$ and $c \to \infty$. We obtain the following feedback gains

$$g_1 = 6/\tau^2 \qquad g_2 = 4/\tau \tag{3.20}$$

Remark: Using the definition of x (i.e., $x_1 = y$ and $x_2 = \dot{y}$), we obtain from Eqs. (3.20) and from the fact that $\lambda = y/(V_c\tau)$ and $\dot{\lambda} = (y + \dot{y}\tau)/V_c\tau^2$ that $u = -V_c(4\dot{\lambda} + 2\lambda/\tau)$. It has the structure of PN with a navigation constant $N' = 4$

and an additional feedback from the LOS angle. Thus, given the LOS angle and its rate, perfect rendezvous can be easily mechanized.

The corresponding equations of motion become

$$\dot{x} = x_2$$
$$\dot{x}_2 = -(g_1 x_1 + g_2 x_2) = \frac{-6}{\tau^2} x_1 + \frac{-4}{\tau} x_2 \tag{3.21}$$

We again guess at a solution of the form

$$x_1 = c_1 \tau^{r_1} + c_2 \tau^{r_2} \qquad x_2 = -c_1 r_1 \tau^{r_1 - 1} - c_2 r_2 \tau^{r_2 - 1}$$

and we substitute it into Eqs. (3.21). However, in this case we obtain $r_1 = 2$ and $r_2 = 3$ and therefore the closed-form expression for the trajectories are

$$x_1 = c_1 \tau^2 + c_2 \tau^3 \qquad x_2 = -2c_1 \tau - 3c_2 \tau^2 \tag{3.22}$$

where the constants c_1 and c_2 can be determined from the initial conditions as follows:

$$c_1 = \frac{3}{t_f^2} x_1(t_0) + \frac{1}{t_f} x_2(t_0)$$
$$c_2 = \frac{-2}{t_f^3} x_1(t_0) - \frac{1}{t_f^2} x_2(t_0) \tag{3.23}$$

Note that, in this case, $x_1(t_f) = 0$ and $x_2(t_f) = 0$ regardless of the initial conditions.

IV. Maneuvering Target

A. Ideal Pursuer Case: Augmented Proportional Navigation (APN)

Rewriting Eq. (3.2)

$$\dot{x} = Ax + Bu + Dw$$
$$A = \begin{bmatrix} 0 & 1 \\ 0 & 0 \end{bmatrix} \qquad B = \begin{bmatrix} 0 \\ 1 \end{bmatrix} \qquad D = \begin{bmatrix} 0 \\ 1 \end{bmatrix} \tag{3.24}$$

We now assume that the target is maneuvering with constant normal acceleration $w = w_0$, starting at t_0. To simplify the discussion we begin with the case $c = 0$ (i.e., no velocity constraints). The case with terminal velocity constraints will be studied later (Sec. IV.D). The cost is therefore

$$J = \frac{1}{2} b x_1^2(t_f) + \frac{1}{2} \int_{t_0}^{t_f} u^2(t)\, \mathrm{d}t \tag{3.25}$$

The first-order necessary conditions are as before

$$\dot{\lambda}_1 = 0 \qquad \dot{\lambda}_2 = -\lambda_1 \qquad u = -\lambda_2 \tag{3.26}$$

The only change is in the terminal conditions, where (since $c = 0$) we get

$$\lambda_1(t_f) = b x_1(t_f) \qquad \lambda_2(t_f) = 0 \tag{3.27}$$

Thus

$$\lambda_1 = bx_1(t_f) \qquad \lambda_2 = bx_1(t_f)(t_f - t) \tag{3.28}$$

Substituting $u = -bx_1(t_f)(t_f - t)$ into Eqs. (3.24) and integrating it from t_0 to t, we get the following expression for $x_1(t)$:

$$\begin{aligned} x_1(t) = &-\frac{b}{6}x_1(t_f)(t_f - t)^3 + \frac{w_0}{2}(t_f - t)^2 + x_2(t_0)(t - t_0) \\ &- \left[\frac{b}{2}x_1(t_f)(t_f - t_0)^2 - w_0(t_f - t_0)\right](t - t_0) + x_1(t_0) \\ &+ \frac{b}{6}x_1(t_f)(t_f - t_0)^3 - \frac{w_0}{2}(t_f - t_0)^2 \end{aligned} \tag{3.29}$$

At $t = t_f$ we get

$$x_1(t_f) = x_1(t_0) + x_2(t_0)(t_f - t_0) + \frac{w_0}{2}(t_f - t_0)^2 - \frac{b}{3}x_1(t_f)(t_f - t_0)^3 \tag{3.30}$$

Solving for $x_1(t_f)$ we obtain

$$x_1(t_f) = \frac{1/b}{(1/b) + (1/3)(t_f - t_0)^3}\left[x_1(t_0) + x_2(t_0)(t_f - t_0) + \frac{w_0}{2}(t_f - t_0)^2\right] \tag{3.31}$$

Introducing this value of $x_1(t_f)$ in $u = -bx_1(t_f)(t_f - t)$, we get a closed-form expression for $u(t)$ for $t_0 \le t < t_f$ from which we use only the initial value of the control as follows:

$$u(t_0) = \frac{-1}{(1/b) + (1/3)(t_f - t_0)^3}\left[x_1(t_0)(t_f - t_0) + x_2(t_0)(t_f - t_0)^2 + \frac{w_0}{2}(t_f - t_0)^3\right] \tag{3.32}$$

Since the procedure for the calculation of Eqs. (3.30–3.32) may be repeated for any $t_0 < t_f$, the subscript 0 may be omitted from t_0 to obtain the closed-loop control:

$$u(t) = \frac{-1}{(1/b) + (1/3)(t_f - t)^3}\left[x_1(t)(t_f - t) + x_2(t)(t_f - t)^2 + \frac{w_0}{2}(t_f - t)^3\right] \tag{3.33}$$

Note that this result is consistent (for no target acceleration) with Eqs. (3.12a). For the special case $b \to \infty$ (i.e., zero miss) we get

$$u(t) = \frac{-3}{(t_f - t)^2}\left[x_1(t) + x_2(t)(t_f - t) + \frac{w_0}{2}(t_f - t)^2\right] \tag{3.34}$$

The derivation may be carried out automatically by MAPLE as presented in Listing 3.3.

Listing 3.3 Gains for ideal pursuer and maneuvering target

```
> #listing 3.3
> # l1(t) is constant and equals b*x1f.
> # the following is the remaining costate equation for l2(t), namely d(l2)/dt=−l1.
> #
> sys:=diff(l2(t), t)+b*x1f;
> dsolve({sys, l2(tf)=0},l2(t));
> assign(");
> #
> # the following is the state equation for x2(t), namely d(x2)/dt=u+w0=−l2+w0.
> #
> sys1:=diff(x2(t),t)+l2(t)−w0;
> dsolve({sys1,x2(0)=v0},x2(t));
> assign(");
> #
> # the following is the state equation of x1(t), namely d(x1)/dt=x2.
> #
> sys3:=diff(x1(t),t)−x2(t);
> dsolve({sys3, x1(0)=x0},x1(t));
> assign(");
> #
> # the terminal conditions for x1(t) is as follows
> #
> x1(t);
> y1:=subs(t=tf,");
> #
> # solving for explicit expressions for x1f
> #
> sys5:={x1f−y1};
> solve(",x1f);
> assign(");
> #
> # evaluating the control u=−l2 at t=0
> #
> u:=−eval(l2(t));
> u0:=subs(t=0,");
> simplify(");
> # limit value for perfect intercept
> #
> limit(",b=infinity);
>
```

Remarks:

1) An alternative method of derivation employs the following Riccati equations (see Chapter 2):

$$-\dot{P} = PA + A^T P - PBB^T P \qquad P(t_f) = \begin{bmatrix} b & 0 \\ 0 & 0 \end{bmatrix}$$

$$\dot{\Theta} = (PBB^T - A^T)\Theta - PDw_0 \qquad \Theta(t_f) = 0$$

where the control is given by $u(t) = -B^T(Px + \Theta)$.

2) As shown previously in Sec. III.A, the solution cannot have a conjugate point. We have, therefore, satisfied the sufficient conditions for optimality.

3) Consider Eqs. (3.33) and (3.34). The latter has a singular point at $t = t_f$ that demands infinite acceleration—unless x_1 and x_2 are both strictly zeros. Equation (3.33), on the other hand, has no such shortcoming. Thus, avoiding the zero-miss requirement (i.e., using finite b) prevents infinite acceleration requirements at interception. Moreover, even when x_1 and x_2 are both strictly zeros, we obtain from Eq. (3.33) that

$$\lim_{b\to\infty} \left[\lim_{t\to t_f} (u(t)) \right] = 0 \tag{3.35}$$

Thus for any finite b (no matter how large) the terminal acceleration vanishes. However, for $x_1 = x_2 = 0$, Eq. (3.34) entails

$$\lim_{t\to t_f} \left[\lim_{b\to\infty} (u(t)) \right] = -\frac{3}{2} w_0 \tag{3.36}$$

This result is not surprising since $t = t_f$ is a singular point at $b \to \infty$.

4) Equation (3.34) can be written as

$$u = -N'\left(V_c \dot{\lambda} + \tfrac{1}{2} w_0\right) \qquad N' = 3 \tag{3.37}$$

a strategy known as APN. Note that to use this strategy the pursuer must measure (or at least estimate) the target's acceleration.

We have, therefore, presented two guidance laws (PN and APN) that are derived in Ref. 2 by other means. We shall now develop the MEL that can also be found in the preceding reference by using the optimality conditions in state space.

B. Nonideal Pursuer Case: MEL

As suggested in Chapter 1, we now assume that the pursuer controls its normal acceleration via a first-order lag, that is,

$$\frac{\ddot{Y}_p(s)}{u(s)} = \frac{1}{Ts + 1} \tag{3.38}$$

where T is the time constant representing the pursuer's acceleration control system and u is the acceleration command. Writing the governing equations in state space, we get

$$\dot{x} = Ax + Bu + Dw$$

$$A = \begin{bmatrix} 0 & 1 & 0 \\ 0 & 0 & 1 \\ 0 & 0 & -\dfrac{1}{T} \end{bmatrix} \qquad B = \begin{bmatrix} 0 \\ 0 \\ \dfrac{1}{T} \end{bmatrix} \qquad D = \begin{bmatrix} 0 \\ 1 \\ 0 \end{bmatrix} \tag{3.39}$$

where $x_3 \equiv \ddot{Y}_p$.

The cost is as before

$$J = \frac{1}{2} b x_1^2(t_f) + \frac{1}{2} \int_{t_0}^{t_f} u^2(t)\,\mathrm{d}t \tag{3.40}$$

We obtain the necessary conditions in the following form

$$\dot{\lambda}_1 = 0 \qquad \dot{\lambda}_2 = -\lambda_1 \qquad \dot{\lambda}_3 = -\lambda_2 + \frac{\lambda_3}{T} \qquad u = -\frac{1}{T}\lambda_3 \tag{3.41}$$

with the terminal conditions

$$\lambda_1(t_f) = b x_1(t_f) \qquad \lambda_2(t_f) = 0 \qquad \lambda_3(t_f) = 0 \tag{3.42}$$

Integrating these equations yields

$$\begin{aligned} \lambda_1 &= b x_1(t_f) \\ \lambda_2 &= b x_1(t_f)(t_f - t) \\ \lambda_3 &= b x_1(t_f)\left[(t_f - t)T - T^2 + T^2 e^{(t-t_f)/T}\right] \end{aligned} \tag{3.43}$$

We conclude that

$$u = -b x_1(t_f)\left[(t_f - t) - T + T e^{(t-t_f)/T}\right] \tag{3.44}$$

Basically we can repeat the process described by Eqs. (3.30–3.35). First we substitute Eq. (3.44) into Eqs. (3.39), and then integrate the resulting differential equations to get an expression for $x_1(t)$. We evaluate this result for $t = t_f$ to obtain

$$\begin{aligned} x_1(t_f) = {} & -\tfrac{1}{3} b x_1(t_f)\hat{t}_f^3 - \tfrac{1}{2} T^3 b x_1(t_f) + 2T^2 b x_1(t_f)\hat{t}_f e^{-(\hat{t}_f/T)} \\ & + \tfrac{1}{2} T^3 b x_1(t_f) e^{-2(\hat{t}_f/T)} + T^2 x_3(t_0) e^{-(t_f/T)} + \tfrac{1}{2} w_0 \hat{t}_f^2 + b x_1(t_f) T \hat{t}_f^2 \\ & + T x_3(t_0)\hat{t}_f + x_2(t_0)\hat{t}_f - T^2 c x_1(t_f)\hat{t}_f - T^2 x_3(t_0) + x_1(t_0) \end{aligned} \tag{3.45}$$

where

$$\hat{t}_f = t_f - t_0 \tag{3.46}$$

Solving for $x_1(t_f)$ yields

$$x_1(t_f) = \frac{\left(T^2 e^{-(\hat{t}_f/T)} + T\hat{t}_f - T^2\right)x_3(t_0) + \frac{1}{2}w_0\hat{t}_f^2 + x_2(t_0)\hat{t}_f + x_1(t_0)}{1 + \frac{1}{3}b\hat{t}_f^3 + \frac{1}{2}T^3 b - 2T^2 b\hat{t}_f e^{-(\hat{t}_f/T)} - \frac{1}{2}T^3 b e^{-2(\hat{t}_f/T)} - bT\hat{t}_f^2 + T^2 b\hat{t}_f} \tag{3.47}$$

The derivation can be performed by MAPLE to facilitate the computations (see Listing 3.4). Alternatively, the derivation can be performed by using the associated Riccati equation.

Substituting Eq. (3.46) into Eq. (3.44), we finally obtain the following expression for u:

$$u(t) = -(g_1 x_1 + g_2 x_2 + g_3 x_3 + g_4 w_0).$$

$$\begin{aligned}
g_1 &= \frac{1}{\tau^2}\frac{6h^2(h - 1 + e^{-h})}{(6/T^3 b) + (2h^3 + 3 - 6h^2 + 6h - 12he^{-h} - 3e^{-2h})} \\
g_2 &= \frac{1}{\tau}\frac{6h^2(h - 1 + e^{-h})}{(6/T^3 b) + (2h^3 + 3 - 6h^2 + 6h - 12he^{-h} - 3e^{-2h})} \\
g_3 &= \frac{6(h - 1 + e^{-h})^2}{(6/T^3 b) + (2h^3 + 3 - 6h^2 + 6h - 12he^{-h} - 3e^{-2h})} \\
g_4 &= \frac{3h^2(h - 1 + e^{-h})}{(6/T^3 b) + (2h^3 + 3 - 6h^2 + 6h - 12he^{-h} - 3e^{-2h})}
\end{aligned} \tag{3.48}$$

where $\tau = t_f - t$ and $h = \tau/T$.

As we have explained, having bounded feedback gains is equivalent to the nonexistence of conjugate points. Investigating the denominator of the g_i reveals that it is strictly positive for all positive h. Thus we have no conjugate point, and the solution is both realizable and optimal.

Finally, we may examine two limit cases. It is not difficult to show that when $T \to 0$, Eqs. (3.48) yields Eq. (3.33). The case $b \to \infty$ (i.e., zero miss) will be referred to as MEL (see Chapter 1) and is easily obtained from Eqs. (3.48).

C. Nonideal Pursuer and Evader Case: SGL

For the sake of completeness we assume now that the evader also controls its normal acceleration $\ddot{Y}_e$ via a first-order lag; that is,

$$\frac{\ddot{Y}_e(s)}{w(s)} = \frac{1}{\Theta s + 1} \tag{3.49}$$

Listing 3.4 Gains for nonideal pursuer and maneuvering target

```
> #listing 3.4
> # l1 is constant and equals b*x1f
> # l2 is linear and equals b*x1f*(tf−t)
> # the following is the remaining costate equation for l3, namely
> dl3/dt=l3/T−l2
> #
> sys:=diff(l3(t),t)+b*x1f*(tf−t)−l3(t)/T;
> dsolve({sys, l3(tf)=0},l3(t));
> assign(");
> simplify(");
> #
> # state equation for x3, dx3/dt=−x3/T+u; x3(0)=a0
> #
> sys1:=diff(x3(t),t)+x3(t)/T+l3(t)/T^2;
> dsolve({sys1, x3(0)=a0},x3(t));
> assign(");
> simplify(");
> #
> # state equation for x2, dx2/dt=x3+w0; x2(0)=v0
> #
> sys2:=diff(x2(t),t)−x3(t)−w0;
> dsolve({sys2,x2(0)=v0},x2(t));
> assign(");
> #
> # state equation for x1, dx1/dt=x2, x1(0)=x0
> #
> sys3:=diff(x1(t),t)−x2(t);
> dsolve({sys3,x1(0)=x0},x1(t));
> assign(");
> #
> # evaluating and solving for x1f
> #
> x1(t);
> subs(t=tf,");
> sys5:=x1f−";
> solve(",{x1f});
> assign(");
> #
> # evaluating the control at t=0
> #
> l3t:=eval(l3(t));
> simplify(");
> u:=−1/T*(");
> u0:=subs(t=0,");
```

where Θ is the time constant representing the evader's acceleration control system and w is the acceleration command. Writing the governing equations in state-space form as in Chapter 1, we get

$$\dot{x} = Ax + Bu + Dw$$

$$A = \begin{bmatrix} 0 & 1 & 0 & 0 \\ 0 & 0 & 1 & 1 \\ 0 & 0 & -\dfrac{1}{T} & 0 \\ 0 & 0 & 0 & -\dfrac{1}{\Theta} \end{bmatrix} \qquad B = \begin{bmatrix} 0 \\ 0 \\ \dfrac{1}{T} \\ 0 \end{bmatrix} \qquad D = \begin{bmatrix} 0 \\ 0 \\ 0 \\ \dfrac{1}{\Theta} \end{bmatrix} \tag{3.50}$$

where the new variable is defined by $x_4 \equiv \ddot{Y}_e$.

The cost is as before

$$J = \frac{1}{2} b x_1^2(t_f) + \frac{1}{2}\int_{t_0}^{t_f} u^2(t)\,\mathrm{d}t \tag{3.51}$$

Assuming a step command $w = w_0$ and repeating the process described by Eqs. (3.30–3.35), we get

$$u(t) = -(g_1 x_1 + g_2 x_2 + g_3 x_3 + g_4 x_4 + g_5 w_0)$$

$$g_1 = \frac{1}{\tau^2}\frac{6h^2(h - 1 + e^{-h})}{(6/T^3 b) + (2h^3 + 3 - 6h^2 + 6h - 12he^{-h} - 3e^{-2h})}$$

$$g_2 = \frac{1}{\tau}\frac{6h^2(h - 1 + e^{-h})}{(6/T^3 b) + (2h^3 + 3 - 6h^2 + 6h - 12he^{-h} - 3e^{-2h})}$$

$$g_3 = \frac{6(h - 1 + e^{-h})^2}{(6/T^3 b) + (2h^3 + 3 - 6h^2 + 6h - 12he^{-h} - 3e^{-2h})} \tag{3.52}$$

$$g_4 = \left(\frac{h}{\bar{h}}\right)^2 \frac{6(\bar{h} - 1 + e^{-\bar{h}})(h - 1 + e^{-h})}{(6/T^3 b) + (2h^3 + 3 - 6h^2 + 6h - 12he^{-h} - 3e^{-2h})}$$

$$g_5 = \frac{\left[3h^2 + 6\left(\frac{h}{\bar{h}}\right)^2(-\bar{h} + 1 - e^{-\bar{h}})\right](h - 1 + e^{-h})}{(6/T^3 b) + (2h^3 + 3 - 6h^2 + 6h - 12he^{-h} - 3e^{-2h})}$$

where $h = \tau/T$ and $\bar{h} = \tau/\Theta$. Listing 3.5 is the MAPLE program for the symbolic operations.

Remarks:

1) Note that the first three feedback terms have not been changed. Moreover, if the evader is significantly more agile (i.e., its time constant is much smaller than the pursuer's), then $h/\bar{h} \to 0$, and we obtain exactly Eqs. (3.48).

Listing 3.5 Gains for nonideal pursuer and evader

```
> #listing 3.5
> #l1 is constant and equals c*x1f
> #l2 is linear and equal c*x1f*(tf−t)
> # the following is the remaining costate equation
> #
> sys:=diff (l3(t),t)+c*x1f*(tf−t)−l3(t)/T;
> dsolve({sys, l3(tf)=0}, l3(t));
> assign(");
> simplify(");
> #
> # state equation for x3, namely: dx3/dt=−x3/T+u/T; x3(0)=a0
> #
> sys1:= diff(x3(t),t)+x3(t)/T+l3(t)/T^2;
> dsolve({sys1, x3(0)=a0}, x3(t));
> assign(");
> simplify(");
> #
> # state equation for x4, dx4/dt=−x4/Theta + b/Theta; x4(0)=aT0
> #
> sys0:=diff(x4(t),t)+(x4(t)−b)/theta;
> dsolve({sys0, x4(0)=aT0}, x4(t));
> assign(");
> simplify(");
> #
> # state equation for x2, dx2/dt=x3+x4; x2(0)=v0
> #
> sys2:=diff(x2(t),t)−x3(t)−x4(t);
> dsolve({sys2,x2(0)=v0}, x2(t));
> assign(");
> #
> # state equation for x1, dx1/dt=x2; x1(0)=x0
> #
> sys3:=diff(x1(t),t)−x2(t);
> dsolve({sys3, x1(0)=x0}, x1(t));
> assign(");
> #
> # evaluating and solving for x1f
> #
> x1(t);
> subs(t=tf,");
> sys5:=x1f−";
> solve(",{x1f});
> assign(");
```

Cont.

Listing 3.5 (Continued)

```
> #
> # evaluating the control at t=0
> #
> l3t:=eval(l3(t));
> simplify(");
> u:=-1/T*(");
> u0:=subs(t=0,");
```

2) From the preceding discussion we infer that here, again, there is no conjugate point; hence, the solution is optimal.

3) Notice that, in practice, a w_0-feedback implementation is highly unlikely since this measurement is typically unavailable to the pursuer. However, it can be shown that if $w(t)$ is assumed to be a stationary white noise (namely, the target's acceleration model is a first-order Markov process), then the first four feedback gains of Eqs. (3.52) are the optimal ones.[6] For this reason, we will refer to this guidance law as stochastic guidance law (SGL).

4) Here again the derivation can be performed by using the associated Riccati equation.

D. Intercepting Maneuvering Targets with Velocity Constraints: Augmented Optimal Rendezvous (AOR)

We consider the ideal system:

$$\dot{x} = Ax + Bu + Dw$$

$$A = \begin{bmatrix} 0 & 1 \\ 0 & 0 \end{bmatrix} \qquad B = \begin{bmatrix} 0 \\ 1 \end{bmatrix} \qquad D = \begin{bmatrix} 0 \\ 1 \end{bmatrix} \tag{3.53}$$

with $w = w_0$. We again introduce the terminal velocity term in the cost; that is,

$$J = \frac{b}{2}x_1^2(t_f) + \frac{c}{2}x_2^2(t_f) + \frac{1}{2}\int_{t_0}^{t_f} u^2(t)\,\mathrm{d}t \tag{3.54}$$

The adjoint equations are identical to the stationary target case of Sec. III.A, and therefore

$$\lambda_1 = bx_1(t_f) \qquad \lambda_2 = bx_1(t_f)(t_f - t) + cx_2(t_f) \tag{3.55}$$

hence

$$u = -[bx_1(t_f)(t_f - t) + cx_2(t_f)] \tag{3.56}$$

We now repeat the procedure as before. Substituting Eq. (3.56) into Eqs. (3.53) and integrating it from t_0 to t, we get expressions for $x_1(t)$ and $x_2(t)$ from which, after evaluation at $t = t_f$, we can solve for the unknown terminal values $x_1(t_f)$ and $x_2(t_f)$, and hence for $u(t)$. Listing 3.6 presents the symbolic operations needed

Listing 3.6 Gains for ideal pursuer with velocity constraints and maneuvering target

```
> #listing 3.6
> # l1(t) is constant and equals b*x1f.
> # the following is the remaining costate equation for l2(t), namely d(l2)/dt=−l1.
> #
> sys:=diff(l2(t),t)+b*x1f;
> dsolve({sys, l2(tf)=c*x2f}, l2(t));
> assign(");
> #
> # the following is the state equation for x2(t), namely d(x2)/dt = u+w0 = −l2+w0.
> #
> sys1:= diff(x2(t),t)+l2(t)−w0;;
> dsolve({sys1, x2(0)=v0},x2(t));
> assign(");
> #
> # the following is the state equation of x1(t), namely d(x1)/dt=x2.
> #
> sys3:=diff(x1(t),t)−x2(t);
> dsolve({sys3,x1(0)=x0},x1(t));
> assign(");
> #
> # the terminal conditions for x1(t) and x2(t) are as follows
> #
> x1(t);
> y1:=subs(t=tf,");
> x2(t);
> y2:= subs(t=tf,");
> #
> # solving for explicit expressions for x1f and x2f.
> #
> sys5:={x1f−y1,x2f−y2};
> solve(",{x1f,x2f});
> assign(");
> #
> # evaluating the control u=−l2 at t=0
> #
> u:=−eval(l2(t));
> u0:=subs(t=0,");
> simplify(");
> #
> # limit value for no velocity constraints
> #
> limit(",c=infinity);
> #
> # limit value for perfect intercept
> #
> limit(",b=infinity);
```

to complete the derivation. The resulting guidance law is then $u(t) = -[g_1 x_1(t) + g_2 x_2(t) + g_3 w_0]$, where

$$
\begin{aligned}
g_1 &= \frac{b\tau\left(\frac{1}{2}\tau c + 1\right)}{\tau^4(cb/12) + \tau c + \tau^3(b/3) + 1} \\
g_2 &= \frac{\frac{1}{3}\tau^3 cb + \tau^2 b + c}{\tau^4(cb/12) + \tau c + \tau^3(b/3) + 1} \\
g_3 &= \frac{\frac{1}{2}\tau^3 b + \tau^4(cb/12) + c\tau}{\tau^4(cb/12) + \tau c + \tau^3(b/3) + 1}
\end{aligned}
\tag{3.57}
$$

and where $\tau = t_f - t$.

An interesting special case is perfect rendezvous obtained by letting $b \to \infty$ and $c \to \infty$. We obtain the following feedback gains

$$
g_1 = 6/\tau^2 \qquad g_2 = 4/\tau \qquad g_3 = 1 \tag{3.58}
$$

The first two gains are already familiar. The meaning of $g_3 = 1$ is to cancel out the target acceleration effect by performing exactly the opposite maneuver such that the corresponding equations of motion become

$$
\begin{aligned}
\dot{x}_1 &= x_2 \\
\dot{x}_2 &= (g_1 x_1 + g_2 x_2 + g_3 w_0) + w_0 = \frac{-6}{\tau^2} x_1 + \frac{-4}{\tau} x_2
\end{aligned}
\tag{3.59}
$$

Remarks:

1) Recall that the problem of perfect intercept (i.e., no velocity constraints) for this ideal model lead to APN (Eq. 3.34), where the target acceleration feedback gain was $\frac{3}{2}$. By employing this value, the perfect intercept control does not cancel out the target acceleration effect as opposed to the perfect rendezvous control. Many missiles, designed against stationary targets, use a gravity-bias term to cancel out the effect of gravity—basically an effective target acceleration. Based on the preceding observation, this gravity-bias term is the correct optimal term to obtain perfect rendezvous but is not optimal with respect to perfect intercepts.

2) It is natural to continue the development with the nonideal pursuer intercepting a maneuvering target with terminal velocity constraints. Unfortunately, in this case the explicit symbolic expressions for the gains are too complicated to be given here. For the interested reader, these gains can be obtained by running the code given in Listing 3.7.

V. MATLAB™ Simulations

A. Forward Simulations

In this section a MATLAB-based numerical simulation will be presented by two representative examples. The pursuer for both cases is modeled by a single time constant (i.e., the nonideal pursuer) and the selected guidance law is the MEL [Eqs. (3.48)].

Listing 3.7 Gains for nonideal pursuer with velocity constraints and maneuvering target

```
> #listing 3.7
> # l1(t) is constant and equals b*x1f.
> # l2(t) is linear and equals c*x2f+b*x1f*(tf−t)
> # the following is the remaining costate equation for l3(t), namely
> #d(l3)/dt=−l2+l3/T; l3(tf)=0.
> #
> sys:=diff(l3(t),t)+c*x2f+b*x1f*(tf−t)−l3(t)/T;
> dsolve({sys, l3(tf)=0},l3(t));
> assign(");
> simplify(");
> #
> # the following is the state equation for x3(t), namely d(x3)/dt=(x3+u)/T;
> u=−l3/T; x3(0)=a0.
> #
> sys1:= diff(x3(t),t)+x3(t)/T+l3(t)/T^2;
> dsolve({sys1,x3(0)=a0}, x3(t));
> assign(");
> simplify(");
> #
> # the following is the state equation for x2(t), namely d(x2)/dt=x3+w0;
> x2(0)=v0.
> #
> sys2:=diff(x2(t),t)−x3(t)−w0;
> dsolve({sys2,x2(0)=v0},x2(t));
> assign(");
> #
> # the following is the state equation for x1(t), namely d(x1)/dt=x2; x1(0)=x0
> #
> sys3:=diff(x1(t),t)−x2(t);
> dsolve({sys3,x1(0)=x0},x1(t));
> assign(");
> #
> #evaluating and solving for explicit expressions for x1f and x2f.
> #
> x1(t);
> y1:=subs(t=tf,");
> x2(t);
> y2:=subs(t=tf,");
> sys5:={x1f−y1,x2f−y2};
> solve(",{x1f,x2f});
> simplify(");
> assign(");
> #
```

Cont.

Listing 3.7 (Continued)

```
> # evaluating the control u=−13 at t=0
> #
> 13t:=eval(13(t));
> u:=−1/T*(");
> u0:=subs(t=0,");
```

The evader of Example 3.1 (Listing 3.8) uses a direct control of its normal acceleration (i.e., the ideal evader) to perform a constant evasive maneuver of 3 g. For Example 3.2 (see Listing 3.9) the evader is nonideal, modeled by a single time constant, and its evasive maneuver is random. Other combinations of pursuer and evader models, evasive maneuvers, and guidance laws can be simulated by changing a few command lines (see Appendix C).

The case of nonideal target maneuver, when the target maneuvers are random, requires some extra explanations. In this case, we set the value *THETA* of the evader's time constant as well as the double value of the standard deviation *nt_2sigma* of the target's maneuvers. A white noise sequence (randomized at each integration step) with the appropriate covariance is generated using the MATLAB function *randn*. The listing of the MATLAB forward simulation for this case is given in Listing 3.9. The relative simplicity of the simulation program was achieved due to the most helpful matrix-writing feature of MATLAB. As can be inferred from the previous examples, the pursuer and evader dynamics can be altered by giving the corresponding state-space matrices *A, B*, and *D* (see Appendix C). The evasive maneuver can be also easily changed by using any desirable value for w. For example, $w = 3*g*\sin(2*t)$ will describe a sinusoidal maneuver of 3 g of magnitude and a 2-rad/s frequency. Finally, the guidance law can be changed by using other options for the gain functions, a few of which will be described in the next section.

B. Guidance Gains

We begin by describing the guidance laws for the MEL family as used in the previous examples. The input parameters are as follows.

t. The time-to-go used to compute the gains. Notice that it is called from the program with the parameter *tgo_meas*, which is the time-to-go as known to the pursuer. (It may include bias and scale factor errors, which in the program are set to 0 and 1, respectively.)

T. The pursuer's time constant used to compute the gains. Also here, notice that it is called from the program with *T_des*, which is the design time constant. Thus, it is not necessarily the exact time constant of the pursuer. It is, in fact, the time constant that the guidance engineer models the pursuer with. The possible difference between these two time constants is what gives rise to the robustness issues that are thoroughly discussed in Chapter 6 of this book.

one_over_b. It is the $1/b$ where b is the miss-distance weighting in the cost function.

Listing 3.8 Forward simulation (Example 3.1)

```
%
% Forward Simulation for the case of Nonideal Pursuer & Ideal Evader
% Step in Target Acceleration
% Minimum Effort Law
%
% General Parameters
%
g      = 9.81;
dt     = 0.01;
dtg    = dt;
%
% Pursuit Evasion Parameters
%
tf     = 3;              % final time
Vc     = 300;            % closing velocity
g_max  = 15;             % maximum pursuer maneuver in g
ncg_max = g_max*g;       % maximum pursuer maneuver in m/s^2
T = 0.5;                 % actual pursuer time constant
T_des = 0.5;             % design value for pursuer time constant
%
% State-Space Description
% ***********************
% x    = [relative pos. ;relative veloc.; missile accel];
% u    = pursuer acceleration command
% w    = evader acceleration
% dx/dt = A x + B u + D w
%

A = [0 1 0
     0 0 1
     0 0 -1/T];
B = [0 0 1/T]';
D = [0 1 0]';
%
% guidance law parameters
%
one_over_b = 1e-03;  % penalty on miss x1(tf)^2*b
gam    = Inf;            % one-sided optimal cont. problem ->
                         % w is assumed to be constant


i_trgf  = 1;             % w is assumed to be measured
%
% time-to-go measurement errors
%
tgo_err  = 0;  % zero bias
tgo_scl  = 1;  % unit scale factor
```

Cont.

Listing 3.8 (Continued)

```
%
% Forward Simulation
%****************
%
% Initial Conditions
%
n     = length A);
x     = zeros(n,1);
nt    = -3*g;          % define a 3g
ntd   = nt;            % evader's maneuver
i   = 0;
J_u   = 0;
% Forward Simulation Loop
for t=0:dt:tf,
 i        = i+1;  % counter
 % guidance gains
 tgo_true = tf-t;                    % true time-to-go
 tgo_meas = tgo_true * tgo_scl + tgo_err;  % measured time-to-go
 k       = min_eff_gains(tgo_meas,T_des,one_over_b,gam,i_trgf);
 % pursuer acceleration command
 u = k* [x;ntd]
% pursuer acceleration command : hard limiter
if(u>ncg_max),
 u = ncg_max;
end
if(u<-ncg_max),
 u = -ncg_max;
end
% updating of the integral control effort
J_u = J_u + u^2*dt;
% state-space equations
w       = ntd;
x_dot = A*x + B*u + D*w;
x       = x  + x_ dot * dt;
% store for graphics
t_vec(i)        = t;
y_vec(i)        = x(1);
y_dot_vec(i) = x(2);
nl_vec(i)       = x(3);
ntd_vec(i)      = ntd;
nc_vec(i)       = u;
J_u_vec(i)      = J_u;
end % END OF for t=0:dt:tf,
%
```

Cont.

Listing 3.8 (Continued)

```
% Plots of forward simulation
%
N      = i;
miss  = x(1);
veloc = x(2);
angle = (180/pi)*x(2)/Vc;
J      = J_u_vec(N)/(g*g*tf);
miss_tit = sprintf('    Miss dist [m]= %g',miss);
ang_tit = sprintf('     Interception angle [deg] %g' ,angle);
J_tit     = sprintf('     Integral of u^2/g^2*tf =%g' ,J);
subplot(2,2,1);
plot(t_vec,y_vec);
xlabel('Time [s]');
ylabel('Relative separation [m]');
grid;
title(miss_tit);
subplot(2,2,2);
plot(t_vec,y_dot_vec);
xlabel('Time [s]');
ylabel('Relative velocity [m/s]');
grid;
title(ang_tit);
subplot(223);
plot(t_vec,nc_vec/g,t_vec,nl_vec/g);
axis([0 tf -g_max*1.5 g_max*1.5]);
xlabel('Time [s]');
ylabel('Interceptor acceleration [g]');
title('Acceleration profile');
grid;
subplot(224);
plot(t_vec,J_u_vec/(g^2*tf));
xlabel('Time [s]');
ylabel(J_tit);
title('Control effort');
grid;
```

gam. This parameter is not relevant to the present chapter but the reader is familiar with it from Chapter 2. It is the γ parameter, which when taken to infinity (as in the present chapter), entails a one-sided optimal control problem. Otherwise, a two-sided optimal control problem, namely a differential game with two players, emerges. This issue will be discussed beginning in Chapter 4.

i_trgf. This parameter takes the values 0 or 1. The 0 value corresponds to the assumption that the evader's acceleration is not measured by the pursuer, whereas the 1 value corresponds to the opposite but common (in the guidance theory literature, e.g., Ref. 2) assumption that the evader's acceleration is measured by the pursuer.

The listing of the function *min_eff_gains* is given in Listing 3.10.

Listing 3.9 Forward simulation (Example 3.2)

```
%
% Forward Simulation for the case of Nonideal Pursuer and Ideal Evader
% Random Target Acceleration
% Minimum Effort Guidance Law
%%
% General Parameters
%
g      = 9.81;
dt     = 0.01;
dtg    = dt;
%
% Pursuit Evasion Parameters
%
tf     = 3;                      % final time
Vc     = 300;                    % closing velocity
g_max  = 15;                     % maximum pursuer maneuver in g
ncg_max = g_max*g;               % maximum pursuer maneuver in m/s^2
T      = 0.5;                    % actual pursuer time constant
T_des    = 0.5;                  % design value for pursuer time constant
THETA  = 0.5;                    % evader time constant
randn('seed',1234567);           % seed for randomizing target acceleration
nt_2sigma  = 3;                  % 2-sigma of target maneuver
nt_sigma  = nt_2sigma*g/2;       % 1-sigma of target maneuver
%
% State-Space Description
% ***********************
%x    = [relative pos.; relative veloc.; missile accel; trg. accel.];
%u    = pursuer acceleration command
% w   = evader acceleration
% dx/dt = Ax + Bu + Dw
%
  A = [0 1 0   0
       0 0 1   1
       0 0 -1/T 0
       0 0 0   -1/THETA];
  B = [0 0 1/T 0]';
  D = [0 0 0 1/THETA]';
%
% guidance law parameters
%
one_over_b = 1e-03;  % penalty on miss x1(tf)^2*b
gam         = Inf;    % one-sided optimal cont. problem ->
                       % w is assumed to be constant
i_trgf= 1;            % w is assumed to be measured
```

Cont.

Listing 3.9 (Continued)

```
%
% time-to-go measurement errors
%
tgo_err   = 0;    % zero bias
tgo_scl   = 1;    % unit scale factor
%
% Forward Simulation
%*****************
%
% Initial Conditions
%
n       = length(A);
x       = zeros(n,1);
i        = 0;
J_u     = 0;
% Forward Simulation Loop
for t=0:dt:tf,
i       = i+1; % counter
% guidance gains
tgo_true = tf-t;                              % true time-to-go
tgo_meas = tgo_true * tgo_scl + tgo_err;  % measured time-to-go
k       = min_eff_gains(tgo_meas,T_des,one_over_b,gam,i_trgf);
% random target maneuver
gain_sig  = nt_sigma * sqrt(2*THETA);  % set the random noise intensity
                                          % to get a correct 1-sigma of
                                           % target maneuver
ntd      = gain_sig * randn/sqrt(dt);       % target maneuver
% pursuer acceleration command
 u = k *x;
% pursuer acceleration command : hard limiter
if(u>ncg_max),
 u = ncg_max;
end
if(u<-ncg_max),
 u = −ncg_max;
end
% compute integral control effort
J_u = J_u + u^2*dt;
% state-space equations
w  = ntd;
x_dot = A*x + B*u + D*w;
x    = x + x_dot * dt;
% store for graphics
```

Cont.

Listing 3.9 (Continued)

```
t_vec(i)        = t;
y_vec(i)        = x(1);
y_dot_vec(i) = x(2);
ntd_vec(i)     = x(4);
nl_vec(i)       = x(3);
nc_vec(i)       = u;
J_u_vec(i)      = J_u;
end    % END OF for t=0:dt:tf,
%
% Plots of forward simulation
%
N      = i;
miss   = x(1);
veloc = x(2);
angle = (180/pi)*x(2)/Vc;
J     = J_u_vec(N)/(g*g*tf);
miss_tit = sprintf(' Miss dist [m]= %g' ,miss);
ang_tit = sprintf('  Interception angle [deg] %g' ,angle);
J_tit = sprintf(' Integral of u^2/g^2*tf= %g' ,J);
subplot(221);
plot(t_vec,y_vec);
xlabel('Time [s]');
ylabel('Relative separation [m]');
grid;
ttitle(miss_tit)
subplot(222);
plot(t_vec,ntd_vec/g);
xlabel('Time [s]');
ylabel('Target Maneuver [g]');
grid;
subplot(223);
plot(t_vec,nc_vec/g,t_vec,nl_vec/g);
axis([0 tf -g_max*1.5 g_max*1.5]);
xlabel('Time [s]');
ylabel('Interceptor acceleration [g]');
grid;
subplot(224);
plot(t_vec,J_u_vec/(g^2*tf));
xlabel('Time [s]');
ylabel(J_tit);
grid;
```

Listing 3.10 Gains for the MEL family

```
function [k,NOout] = min_eff_gains(t,T,one_over_b,gam,i_trgf)
%
% k = min_eff_gains(t,T,one_over_b,gam,i_trgf)
%
% Generalized MEL gains for
%
% dx/dt = Ax + Bu + Dw
%
% A = [0 1 0        B = [0 0 1/T]'    D = [0 1 0]'
%      0 0 1
%      0 0 −1/T]
%
% J= x' (tf)*diag([b 0 0])*x(tf) + ||u||2  − gamma^2 * ||w||2 −>saddle point
%
% t          : time-to-go
% T          : pursuer time constant
% one_over_b : 1/b where b is the miss-distance weighting
% gam        : evader maneuver weighting
% i_trgf     : 1 - apply feedback from evader acceleration
%              0 - zero gain on evader acceleration
%
h  = t/T;
T2 = T^2;
T3 = T^3;
h2 = h*h;
h3 = h2*h;
he  = exp(−h);
he2 = exp(−2*h);
if(gam~=Inf),
 gam2i = 1/gam^2;
 f_gam = 1−gam2i;
else
 gam2i = 0;
 f_gam = 1;
end
eps1 = 1e−014;
eps2 = 1e−014;
NOnum = 6.*h2*(he−1.0+h);
NO1  = (6/T2)*(he−1+h);
NO2  = (6*h/T)*(he−1+h);
NOden = 2.0*h3*f_gam+3.0+6.0*h−6.0*h2−12.0*h*he−3.0*he2;
NOden = NOden + eps1 + one_over_b*6/T3;
NTAG  = NOnum/NOden;
NOout = NOden - eps1;
```

Cont.

Listing 3.10 (Continued)

```
KL   = NTAG*(he + h −1)/(h2 + eps2);
if(i_trgf==1),
  k = [−NO1/NOden, −NO2/NOden,
  −KL, −NTAG/2];
  else
  k   = [−NO1/NOden, −NO2/NOden, −KL, 0];
end
return
end
```

Other families of guidance laws can be tested by calling other functions. For example, the AOR family of solutions for the ideal pursuer and evader case (again a generalized version that allows a differential game formulation) are computed by replacing the line

$$k = min\_eff\_gains\,(tgo\_meas,\ T\_des, one\_over\_b,\ gam,\ i\_trgf);$$

by the line

$$k = ideal\_purs\_ev\_gains(tgo\_meas,\ b,\ c,\ gam,\ i\_trgf)$$

which activates the subroutine in Listing 3.11. Here b and c are the weights on miss distance and relative normal velocity at intercept, $gam = Inf$ (i.e., infinity) is the value relevant to the present chapter (otherwise a differential game is obtained; see Chapter 4), and as before, $i\_trgf = 0$ corresponds to unmeasured evader's acceleration whereas *i_trgf* is 1 when it is measured.

A guidance law from the PN family (which in some cases coincides with some of the former guidance laws by taking various parameters b and c) can be selected by the following command:

$$k = prop\_nav\_gains(tgo\_meas,\ NTAG,\ i\_aug)$$

which activates the subroutine in Listing 3.12. Here *NTAG* is the navigation constant and where *i_aug* is 0 for the conventional PN, 1 for APN as it is derived from perfect interception, and 2 for a more commonly used version when "g-bias" correction is used. It differs from the former by the evader's acceleration gain derived from perfect rendezvous.

C. Adjoint Simulations

The adjoint simulation associated with the Example 3.1 is performed by the MATLAB program in Listing 3.13, which should be executed after executing Listing 3.8. The theory by which the adjoint simulations were obtained is explained in Chapter 2. Adjoint simulations of the nonideal pursuer and evader case, where the target maneuvers in a random manner, can be performed by the subroutine given in Listing 3.14, which should be executed after executing Listing 3.9. Although not all possibilities have been exhausted, the examples given are sufficient for getting a good idea of the numerical simulation technique.

Listing 3.11 Gains for the AOR family

```
function k = ideal_purs_ev_gains(t,b,c,gam,i_trgf)
%
% k = ideal_purs_ev_gains(tgo_meas,b,c,gam,i_trgf);
%
% Gains computation for ideal pursuer and evader case
%
% dx/dt = Ax + Bu + Dw
%
% A = [0 1        B = [0 1]'     D = [0 1]'
%      0 0]
%
% J = x' (tf)*diag([b c])*x(tf) + ||u||^2 − gamma^2 * ||w||^2 ->saddle point
%
% t     : time-to-go
% gam   : evader maneuver weighting
% b     : weight on x(tf)
% c     : weight on x_dot(tf)
% gam   : weight on 2-norm of w
% i_trgf : 0 - no gain on target maneuver (i.e., target is smart or unmeasurable)
%          1 - optimal gain on measurable target maneuver
%
h    = t;
h2   = h*h;
h3   = h2*h;
h4   = h3*h;
cb   = c*b;
if(gam ~=Inf),
  gam2 = gam^2;
  fgam = 1-1/gam2;
  fgam2 = fgam^2;
else
  fgam = 1;
  fgam2 = 1;
end
  den = h4 * fgam2 * cb/12 + fgam * (h*c + h3*b/3) + 1;
  k1  = −b*t*(0.5*t*c*fgam + 1) / den;
  k2  = −(h3*c*b*fgam/3 + h2*b + c) / den;
  k3  = −(h4*c*b*fgam/12 + (h3*b/2+c*h) ) / den;
  if(i_trgf==0),
  k3 = 0;
end
k = [k1 k2 k3];
return
end
```

Listing 3.12 Gains for the PN family

```
function k = prop_nav_gains(t,NTAG,i_aug)
%
% k = prop_nav_gains(t,NTAG,i_aug)
%
% Gains for proportional navigation (pn) / augmented pn
%
% t        : time-to-go
%
% NTAG     : navigation constant
%
% i_aug = 0 : proportional navigation     - No target acceleration feedback
% i_aug = 1 : augmented prop. navigation  - Target acceleration feedback
%             derived from optimal interception.
%             This is the classical version of
 augmented prop. navigation
%
% i_aug = 2 : aug. prop. navigation       - Target acceleration feedback
%             derived from optimal rendezvous.
%             The commonly used version of
 augmented prop. navigation
%                for g-bias compensation.
t1  = t + 1e-08;
k1  =
 -NTAG/t1^2;
k2  = -NTAG/t1;
k3  = 0;
if(i_aug==0),
 k4   = 0;
end
if(i_aug==1),
 k4   = -NTAG/2;
end
if(i_aug==2),
 k4   = -1;
end
k = [k1 k2 k3 k4];
return
end
```

VI. Numerical Examples

A. Ideal Pursuer and Evader

In this section the theory for the case of ideal pursuer and evader will be demonstrated via a simple example. In this example, the interception conflict lasts 3 s; the conflict is assumed to be tail chase or head-on and the closing velocity is 300 m/s.

Listing 3.13 Adjoint simulation for Example 3.1

```
%
% Adjoint Simulation
%***************
%
% initial conditions
%
x      = zeros(n,1);
x(1)   = 1;            %
Analyze y(tf)
i      = 0;
z      = 0;
z_rnd = 0;
nt     = −3*g;
N      = tf/dt + 1;
for t=0:dt:tf,
% compute guidance gains
tgo_true = t;                          % true time-to-go
tgo_meas = tgo_true * tgo_scl + tgo_err;
% measured time-to-go
i = i+1;
j = N−i+1;
k = min_eff_gains(tgo_meas,T_des,one_over_b,gam,i_trgf);
k_x   = k(1:3); % state-feedback gain
k_w   = k(4);   % disturbance (feed-forward) gain
% adjoint system's state-space equations
x_dot   = (A+B*k_x)'*x;
x       = x  + x_dot * dt;
z       = z + (D+B*k_w)'*x   * dt;
z_rnd  = z_rnd + ((D+B*k_w)'*x)^2 * dt;
t_vec(i)        = t;
z1_vec(i)       = nt*z;
h1_vec(i)       = x(2);
end % end of for t=0:dt:tf
h1_vec = h1_vec * Vc * 10 * pi/180;
% plots
 subplot(2,1,1);
 plot(t_vec,z1_vec);
 xlabel('Time to go [s]');
 ylabel('Miss distance [m]');
 grid;
 title('Miss-distance due to a 3 g of evader step maneuver');
 subplot(2,1,2);
 plot(t_vec,h1_vec);
 xlabel('Time to go [s]');
 ylabel('Miss distance [m]');
 grid;
 title('Miss-distance due to Heading error of 10 deg.');
```

Listing 3.14 Adjoint Simulation for Example 3.2

```
%
% Adjoint Simulation
%***************
%
% initial conditions
%
x         = zeros(n,1);
x(1)      = 1;              % Analyze y(tf)
i         = 0;
z         = 0;
z_rnd     = 0;
nt        = -3*g;
N         = tf/dt + 1;
for t=0:dt:tf,
% compute guidance gains
tgo_true = t;                          % true time-to-go
tgo_meas = tgo_true * tgo_scl + tgo_err; % measured time-to-go
i = i+1;
j = N-i+1;
k = min_eff_gains(tgo_meas,T_des,one_over_b,gam,i_trgf);
k_x   = k(1:4); % state-feedback gain
k_w   = 0;      % disturbance (feed-forward) gain
% adjoint system's state-space equations
x_dot    = (A+B*k_x)'*x;
x        = x  + x_dot * dt;
z        = z  + (D+B*k_w)'*x  * dt;
z_rnd    = z_rnd + ((D+B*k_w)'*x)^2 * dt;
t_vec(i)          = t;
z_rnd1_vec(i)     = sqrt(z_rnd)*gain_sig;
h1_vec(i)         = x(2);
end % end of for t=0:dt:tf,
h1_vec = h1_vec * Vc * 10 * pi/180;
% plots
subplot(2,1,1);
plot(t_vec,z_rnd1_vec);
xlabel('Time to go [s]');
ylabel('Miss distance [m]');
grid;
title('Miss distance due to a 3 g of random evader maneuver');
subplot(2,1,2);
plot(t_vec,h1_vec);
xlabel('Time to go [s]');
ylabel('Miss distance [m]');
grid;
title('Miss distance due to Heading error of 10 deg.');
```

The guidance laws will be designed using a linearized model of the pursuit-evasion conflict; however, in the simulations a maximum commanded pursuer acceleration of 15 g will be assumed. This will allow for a more realistic comparison of different guidance laws. The effects of five guidance laws will be analyzed.

PN (Proportional navigation). This is the guidance law illustrated in Sec. III.B [Eqs. (3.11)] where we take $b = 10^7$ and $c = 0$. Note that it is identical to the APN guidance law of Sec. IV.A [Eq. (3.33)] when $w_0 = 0$. In cases where this guidance law is applied with nonzero w_0, it can not be considered as optimal.

OR (Optimal rendezvous). This guidance law was discussed in Sec. III.A [Eqs. (3.10)], where we take $b = 10^7$ and $c = 10^7$. Also here the target maneuver is assumed to be zero (i.e., $w_0 = 0$), and thus when it is nonzero, it cannot be considered as optimal.

APN (Augmented proportional navigation). This guidance law is exactly like PN except that the target's acceleration is assumed to be known to the evader. It is discussed in Sec. IV.A, where we apply $b = 10^7$ in Eq. (3.33).

AOR (Augmented optimal rendezvous). This guidance law is discussed in Sec. IV.D, where $b = 10^7$ and $c = 10^7$. It is exactly like OR except that the target acceleration is assumed to be known to the evader.

APNR (Augmented proportional navigation and rendezvous). This commonly used guidance law is a combination of APN and AOR. The gains of APN are used for relative separation and relative velocity, but the gain of AOR is used for target acceleration. As such, it is possible to implement the first two terms in Eq. (3.33) by a simple gain on the LOS rate. Moreover, the gain on the target acceleration, which is identical to that of AOR, ensures near-zero normal separation velocity.

Simulation Results. All five guidance laws were tested via the MATLAB-based simulations on the (forward) dynamics of Eqs. (3.24). To allow for an easy comparison, we classify the results according to their error sources (i.e., target acceleration step and heading error). The results are illustrated by comparing the graphs of relative separation, relative angle (i.e., the quotient of the relative velocity and closing velocity in degrees), commanded pursuer acceleration, and $J_u(t) = \int_0^t u^2(\tau)\,d\tau$. Notice that $J_u(t_f)$ is the pursuer's control effort, which is part of the cost function of Eq. (3.3).

1. Step of 3 g *in Target Acceleration*

The results for this type of error source are depicted in Figs. 3.1–3.4. For the case of a target acceleration step, APNR and AOR guidance laws coincide because they both apply a complete cancellation of the target disturbance via target-acceleration feedback. Indeed, in Figs. 3.1–3.4 the results for these two guidance laws are identical. This cancellation results, as seen in Fig. 3.1, in a zero-relative separation throughout the interception conflict for the APNR/AOR. The price that is paid for this zero-relative separation is, of course, the control cost. It is seen in Fig. 3.4 that the integral-squared commanded acceleration is larger for the APNR/AOR than for the APN, but the maximal required acceleration is higher for the APN, as can be seen in Fig. 3.3.

However, comparing the value of $J_u(t_f)$ in Fig. 3.4 for PN and OR reveals a result that may seem surprising at first sight: The PN control cost is larger than that of OR. Because of the extra constraint it involves, one would expect that the OR

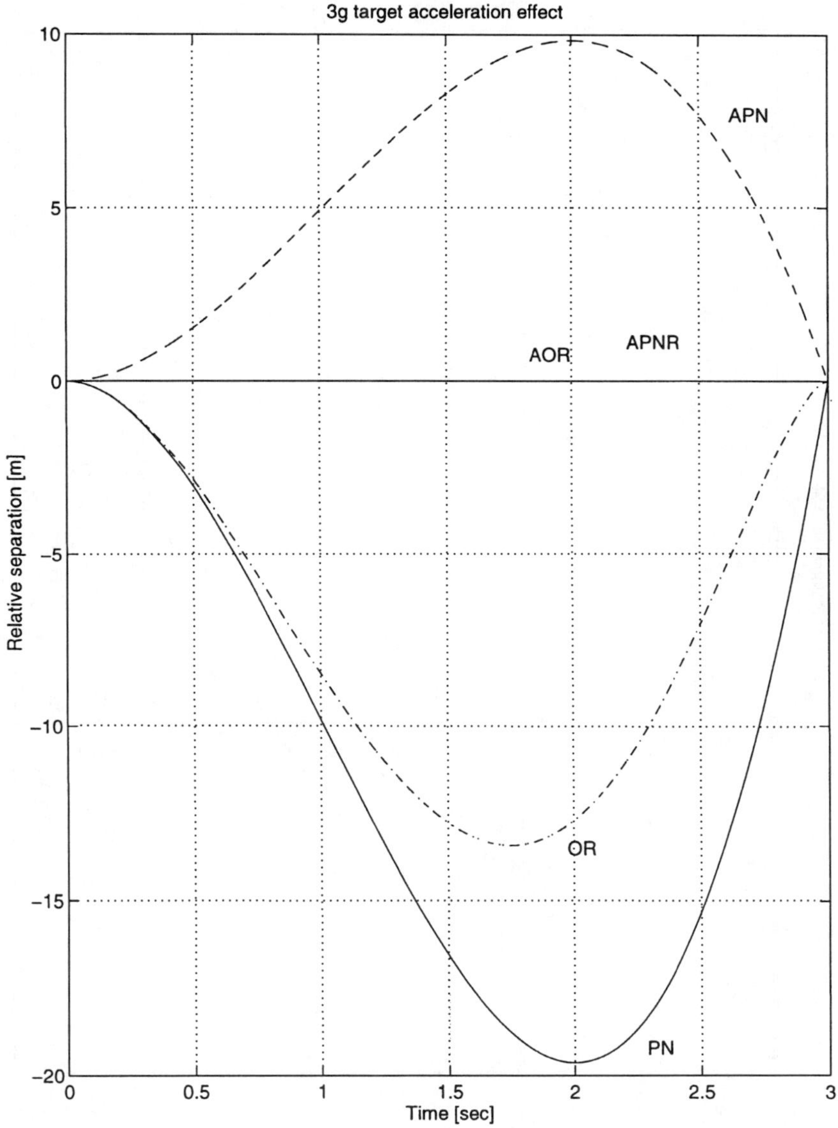

Fig. 3.1 Relative separation due to constant target maneuver.

would have required more control effort. However, we should keep in mind that, for this case, the PN and OR control laws are suboptimal control laws, where we arbitrarily chose to take $w_0 = 0$ (for design purposes) but test these guidance laws for $w_0 \neq 0$. Thus, while the extra end-separation velocity requirement necessarily leads to larger control costs for the optimal guidance law, when a suboptimal guidance law is considered, one cannot tell in advance which control law (PN or OR) will require more control effort.

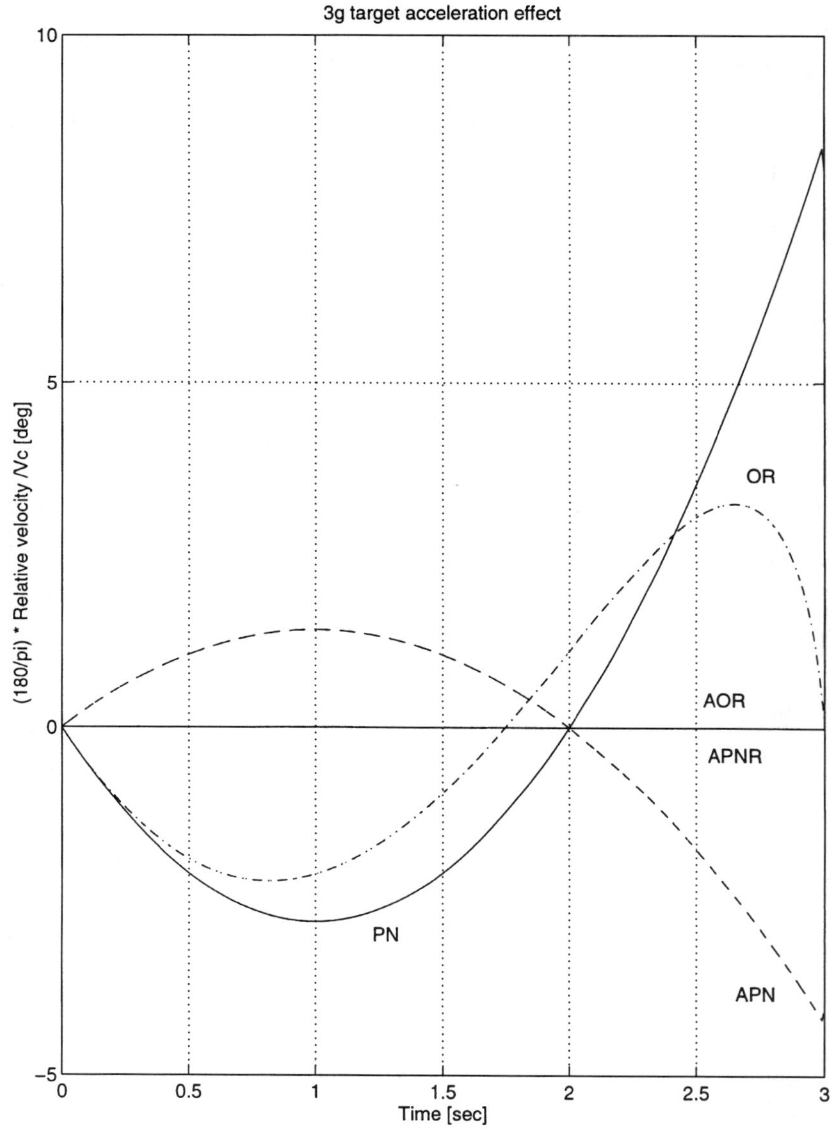

Fig. 3.2 Relative velocity due to constant target maneuver.

2. Heading Error

With an initial heading error only (a 10-deg error in our case), we remain with only two guidance laws, namely PN and OR. Notice that here APN and APNR coincide with PN, while AOR coincides with OR. Moreover, both PN and OR can be considered to be optimal guidance laws with the guarantee that more constraints will result in larger control efforts. It is seen from Fig. 3.5 that both PN and OR show practically zero miss, whereas from Fig. 3.6 we see that OR also exhibits a

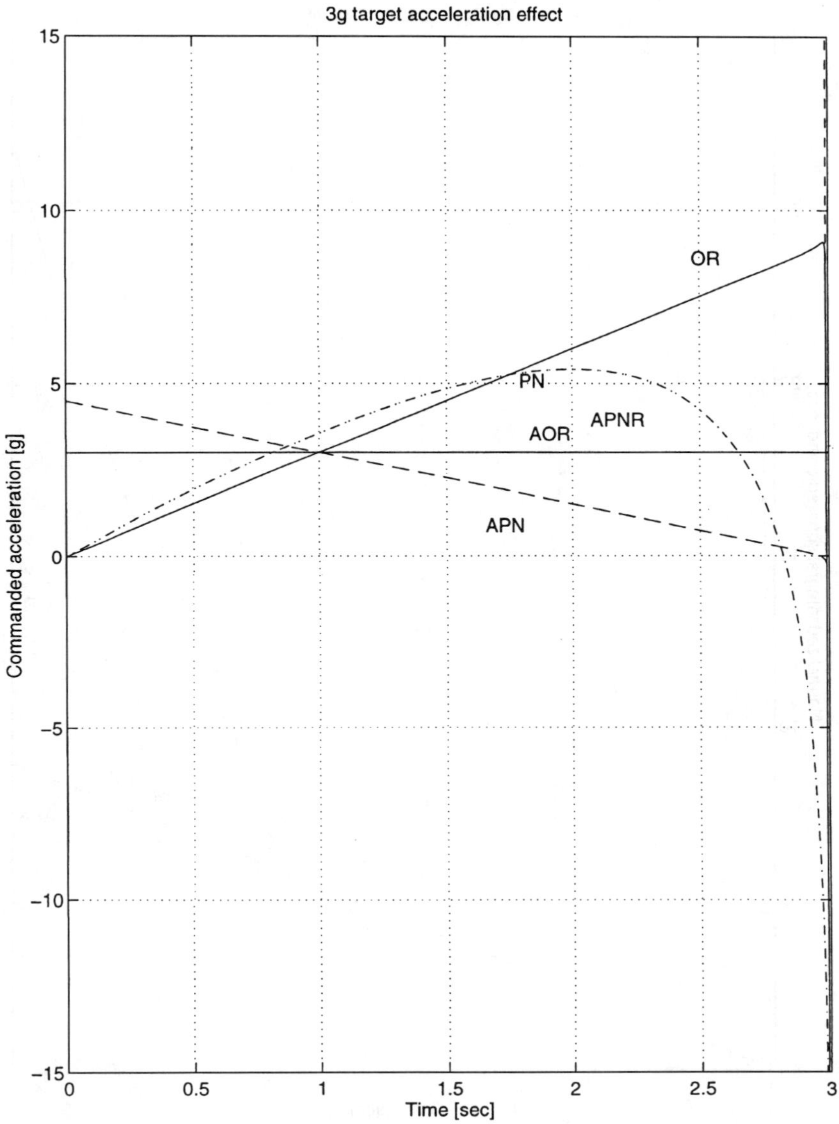

Fig. 3.3 Missile acceleration due to constant target maneuver.

zero-relative velocity. On the other hand, the PN law does not drive the relative end velocity to zero. If this had been the case, it would require more control cost as Figs. 3.7 and 3.8 illustrate. With both PN and OR optimal, the relation between their corresponding control costs is as expected, that is, more constraints require more control.

Remark: For the preceding two cases, the APNR turns out to be the guidance scheme with the minimal required acceleration. This fact may explain the common use of this suboptimal strategy.

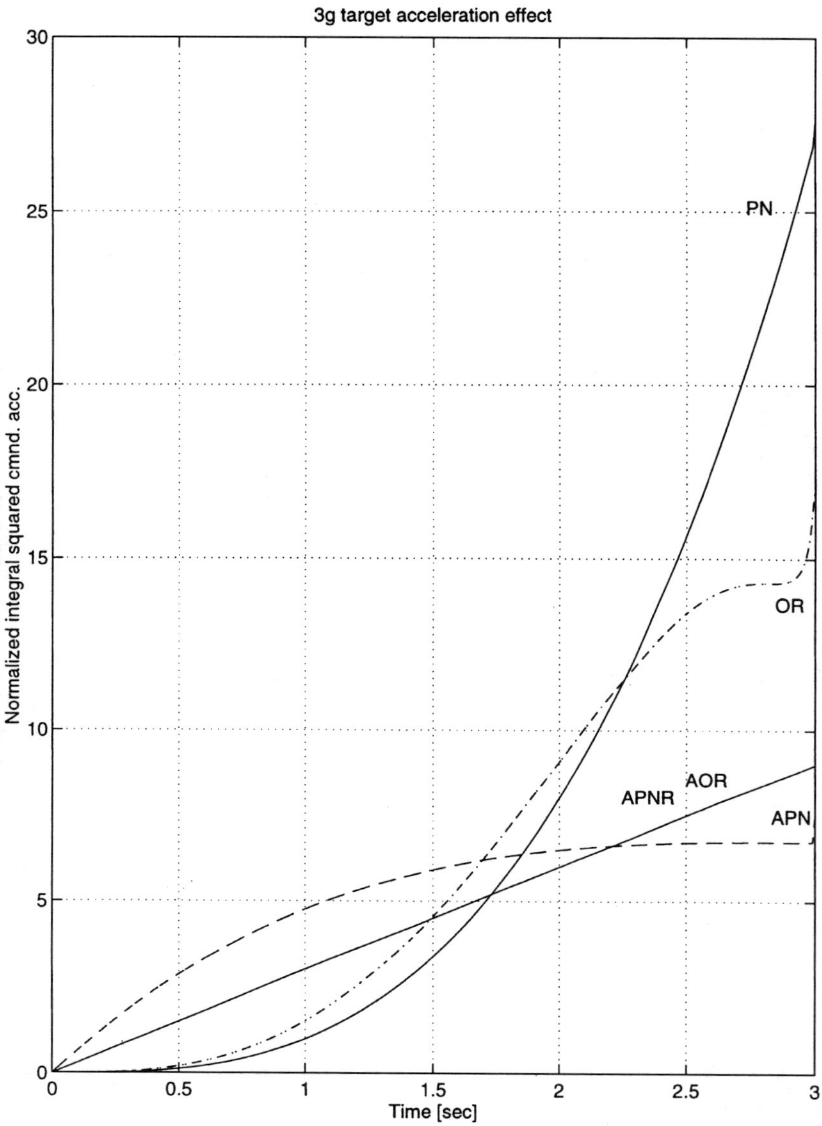

Fig. 3.4 Control effort due to constant target maneuver.

3. *Trajectory Shaping*

For the preceding results we compared various guidance laws that resulted from the guidance law of Sec. IV.A for infinite and zero values of c. However, a continuum of guidance laws exist for all real positive or negative values of c for which a conjugate point does not appear within the pursuit-evasion conflict duration. It should be noted that for positive c no conjugate points can appear even

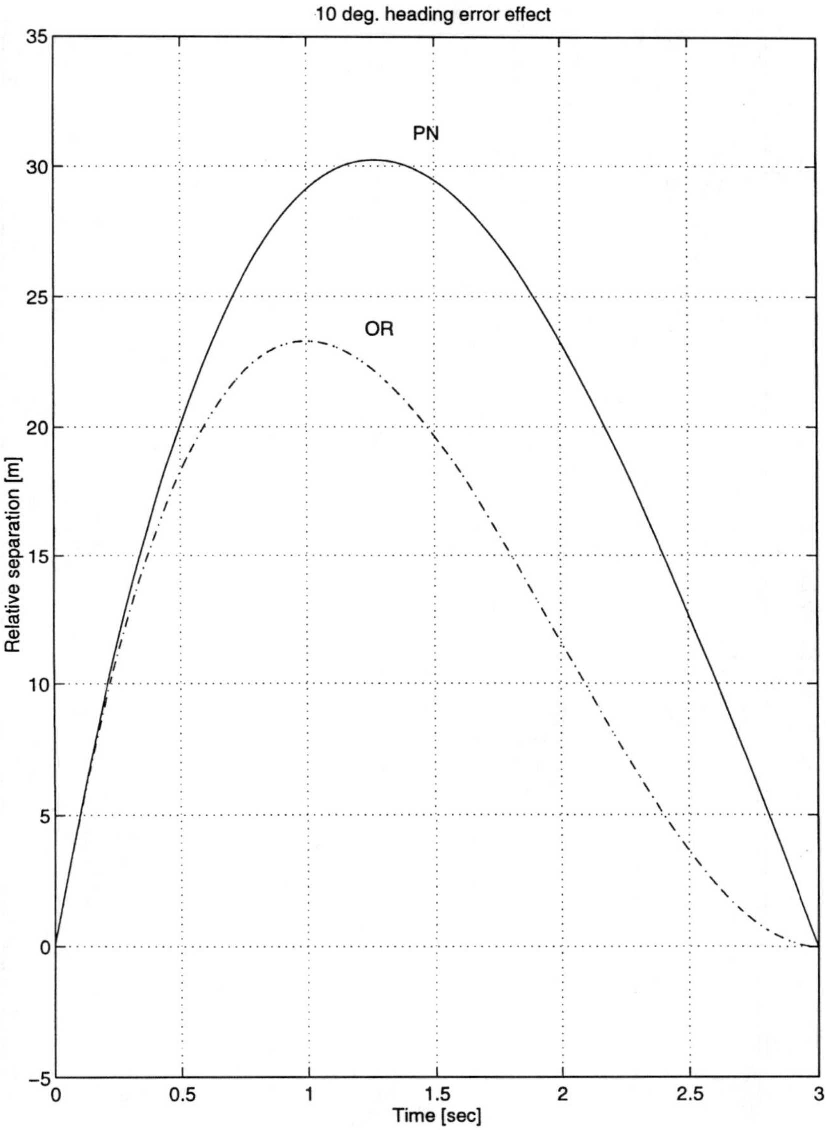

Fig. 3.5 Relative separation due to heading error.

for arbitrarily long conflict duration. However, for a negative c, the conflict duration for which a conjugate point does not appear is limited. We have already shown that a positive infinite c (AOR) leads to zero-relative velocity at interception. We will see now that negative values of c lead to large relative velocities at interception, and that the more negative c is, the larger the angle of interception. All problem data are the same as in Secs. VI.A.6.a and VI.A.6.b, with the exception that we chose to omit the commanded pursuer acceleration limit. Therefore, our analysis is

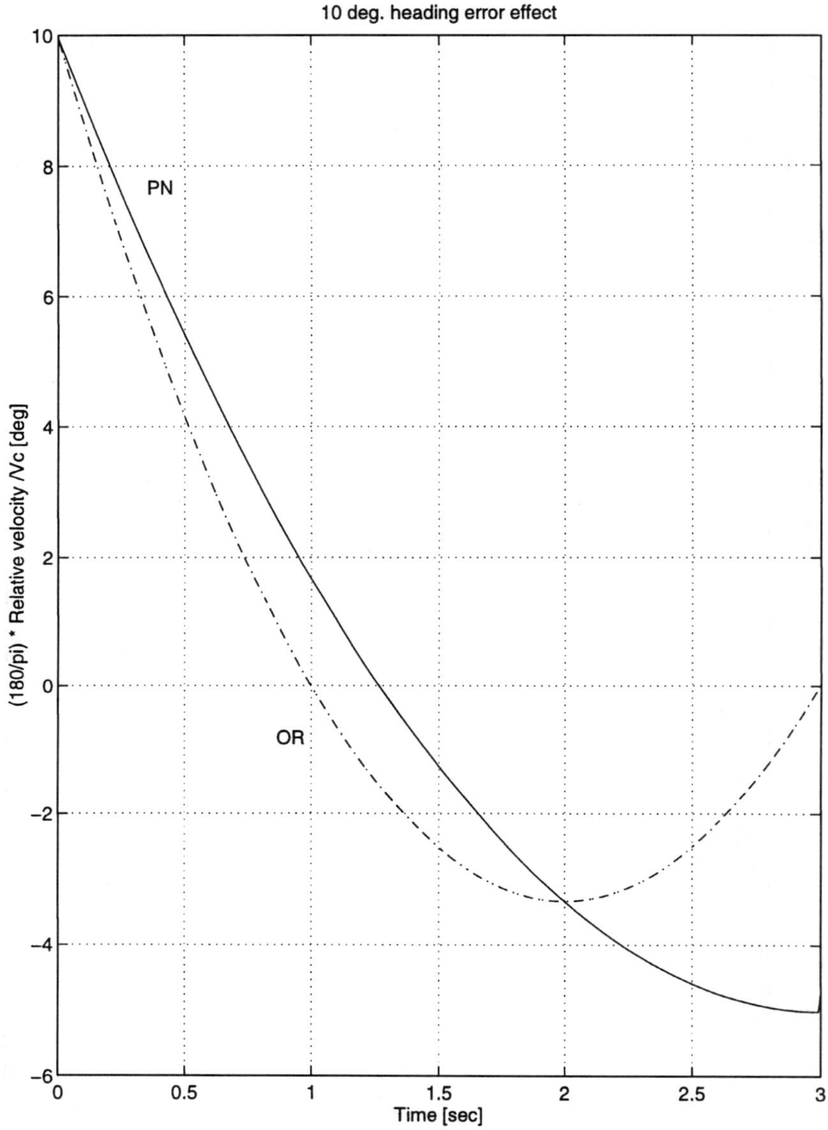

Fig. 3.6 Relative velocity due to heading error.

completely linear and adjoint simulation analysis can also be applied. The guidance laws that we compare differ in their corresponding values of c. We analyze the results for a 10-deg. heading error.

In Fig. 3.9 we see the trajectories for five different values of c: -1.2 and -0.6 (note that for a pursuit-evasion conflict of 3 s the smallest c that avoids the conjugate point is $-4/3 = -1.333$), 0 (corresponds to APN), $+1.2$, and 10^7 (corresponds to AOR). Notice that the more negative c becomes, the interception occurs at a

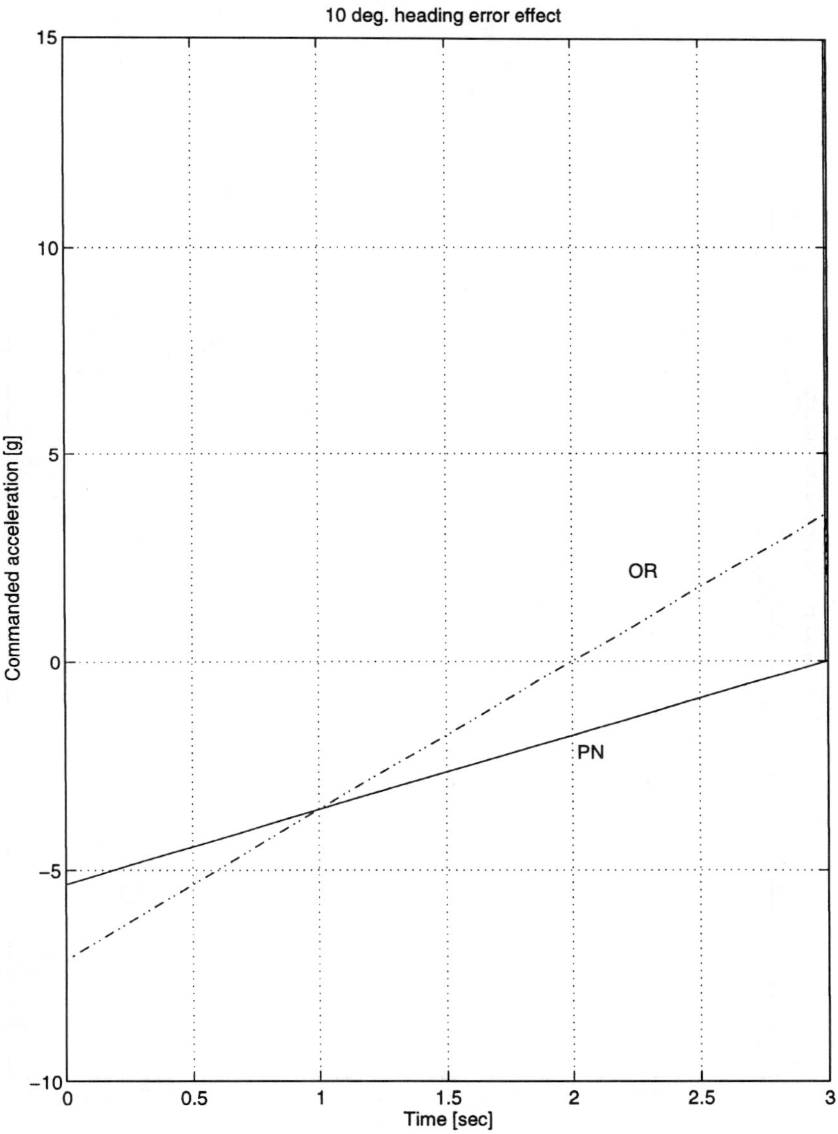

Fig. 3.7 Missile acceleration due to heading error.

steeper angle and the maximum relative separation throughout the conflict is larger. On the other hand, positive values of c tend to lead to flat trajectories with a large c leading to a zero-hit angle.

The results of Fig. 3.9 were all obtained for a pursuit-evasion conflict duration of $t_f = 3$ s. If we are interested in knowing the effect of a 10-deg heading error on the hit angle and the miss distance for various duration times, we can choose between two possibilities:

1) Perform forward simulations for various values of t_f.

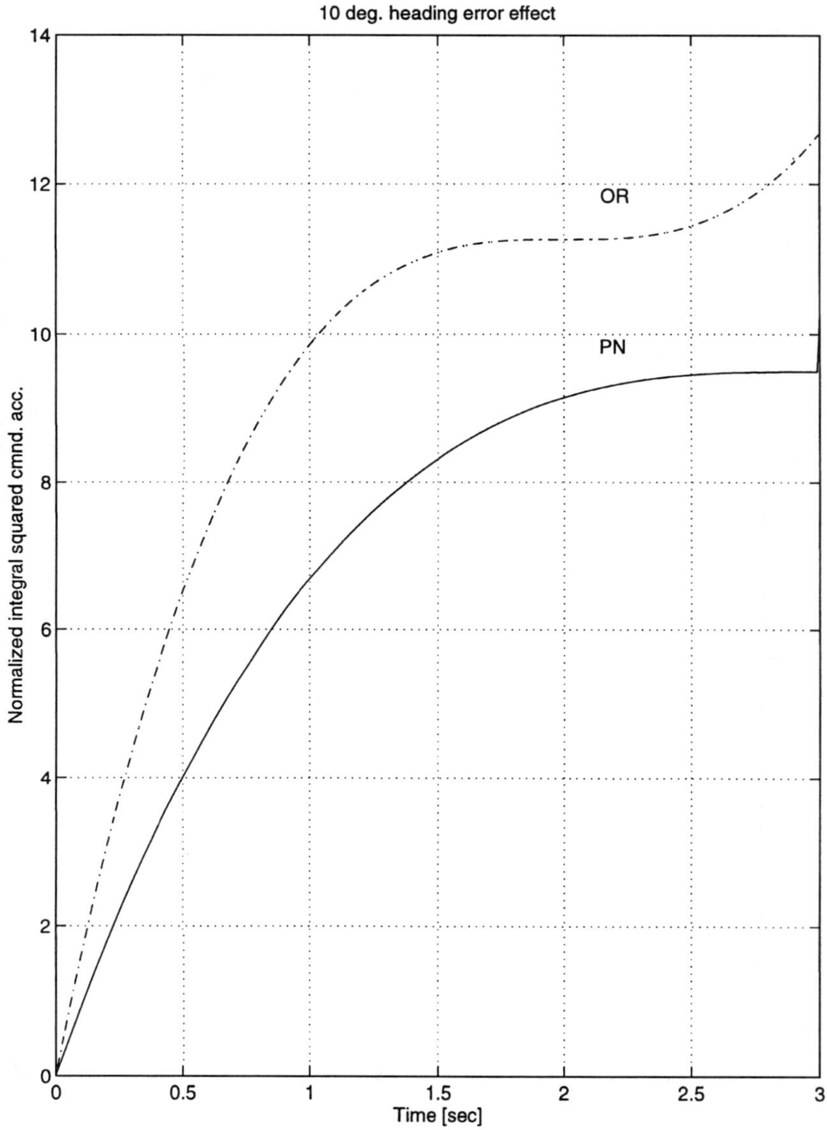

Fig. 3.8 Control effort due to heading error.

2) Perform adjoint system simulation, which yields all the required results in a single run.

Possibility 2 is our choice, for obvious reasons. In Figs. 3.10 and 3.11 we see the results of adjoint simulation. Focus first on Fig. 3.10, the abscissa in this graph is the time-to-go in seconds, and the ordinate is the interception angle in degrees. Each point on a curve reflects the interception angle that is read on the ordinate, where the interception started at a time-to-go that is read from

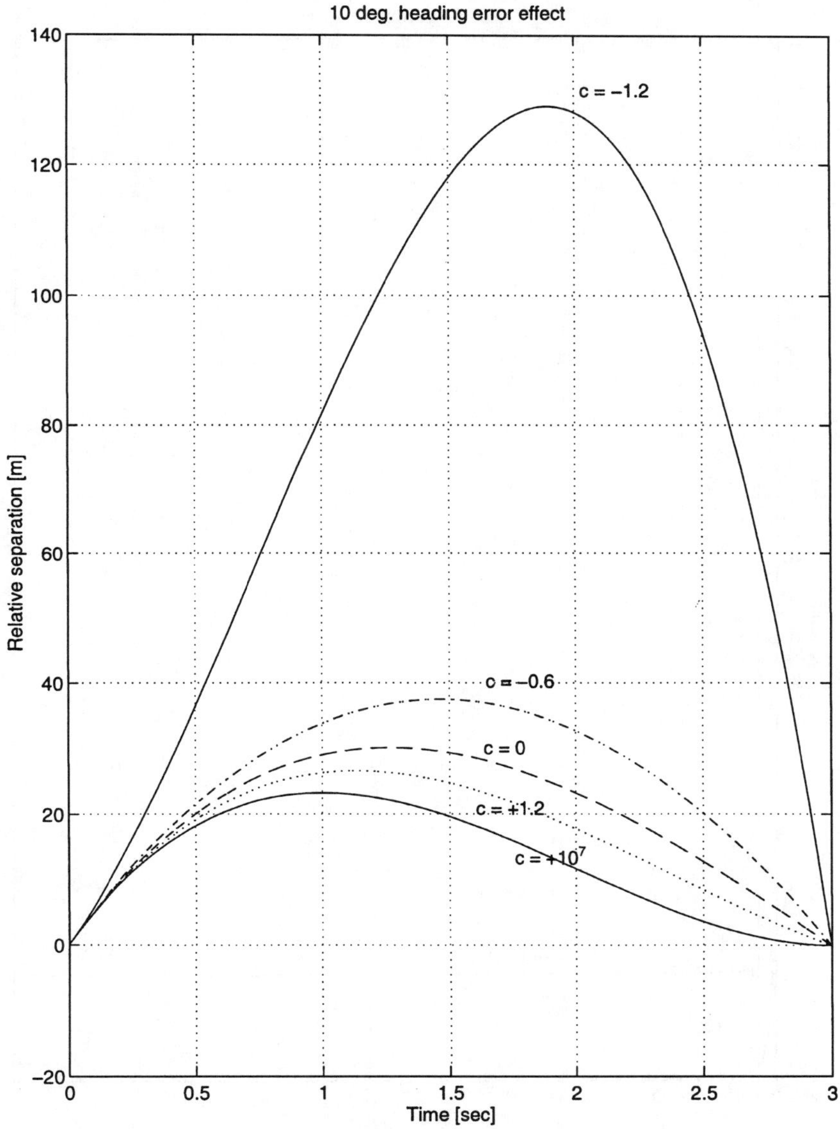

Fig. 3.9 Trajectory shaping.

the abscissa (i.e., conflict duration that is read from the abscissa) with a 10-deg heading error. We see from Fig. 3.10 that for a negative c the interception angle is larger with larger t_f. On the other hand, as c increases, the hit angle gets smaller. Also the larger t_f is, the smaller the hit angle for positive values of c. From Fig. 3.11 we notice that for a large t_f, we have negative values of c resulting in larger miss distances; this is because we took a large but finite $b = 10^7$.

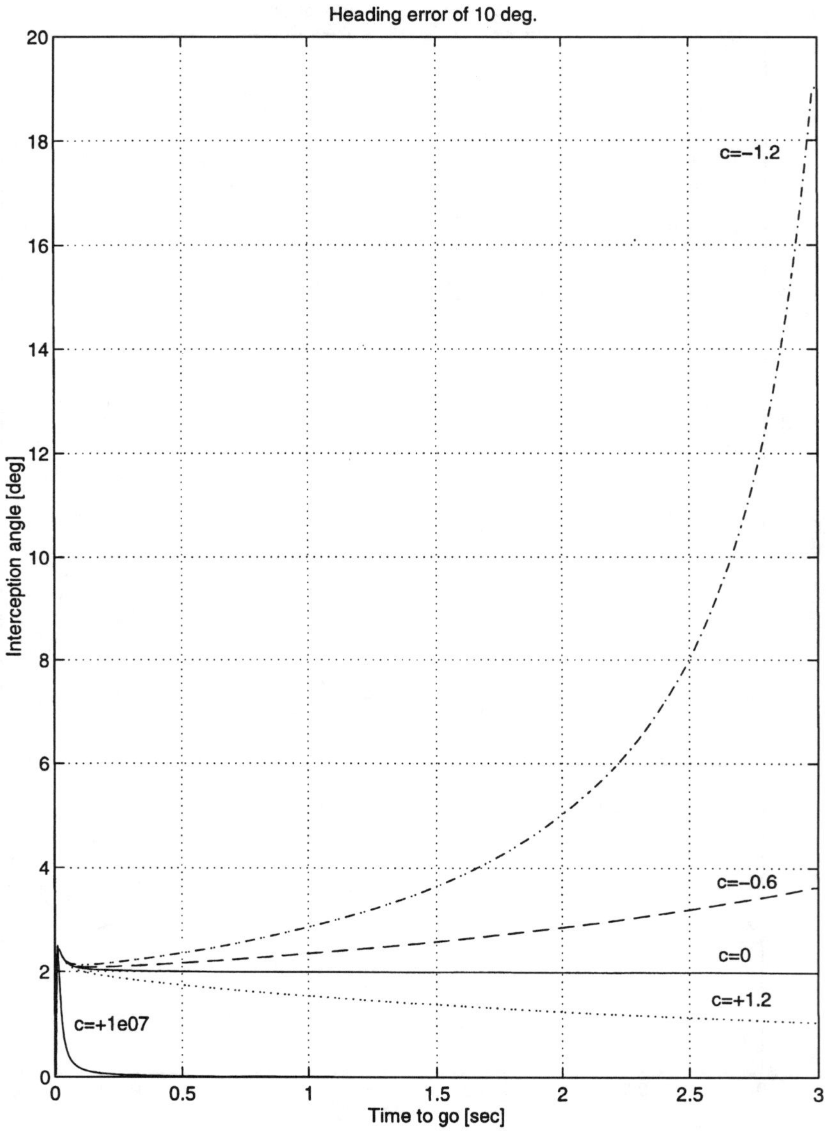

Fig. 3.10 Interception angle with trajectory shaping.

B. Nonideal Pursuer and Ideal Evader—The Effect of Finite Miss-Distance Weighting

In this section the theory for the case of nonideal pursuer, which is modeled by a single time constant, and an ideal evader will be demonstrated. The guidance law of Sec. IV.B is a generalization of the MEL of Zarchan[2] for noninfinite miss-distance weighting b. Whenever this weighting has a finite value, the resulting guidance

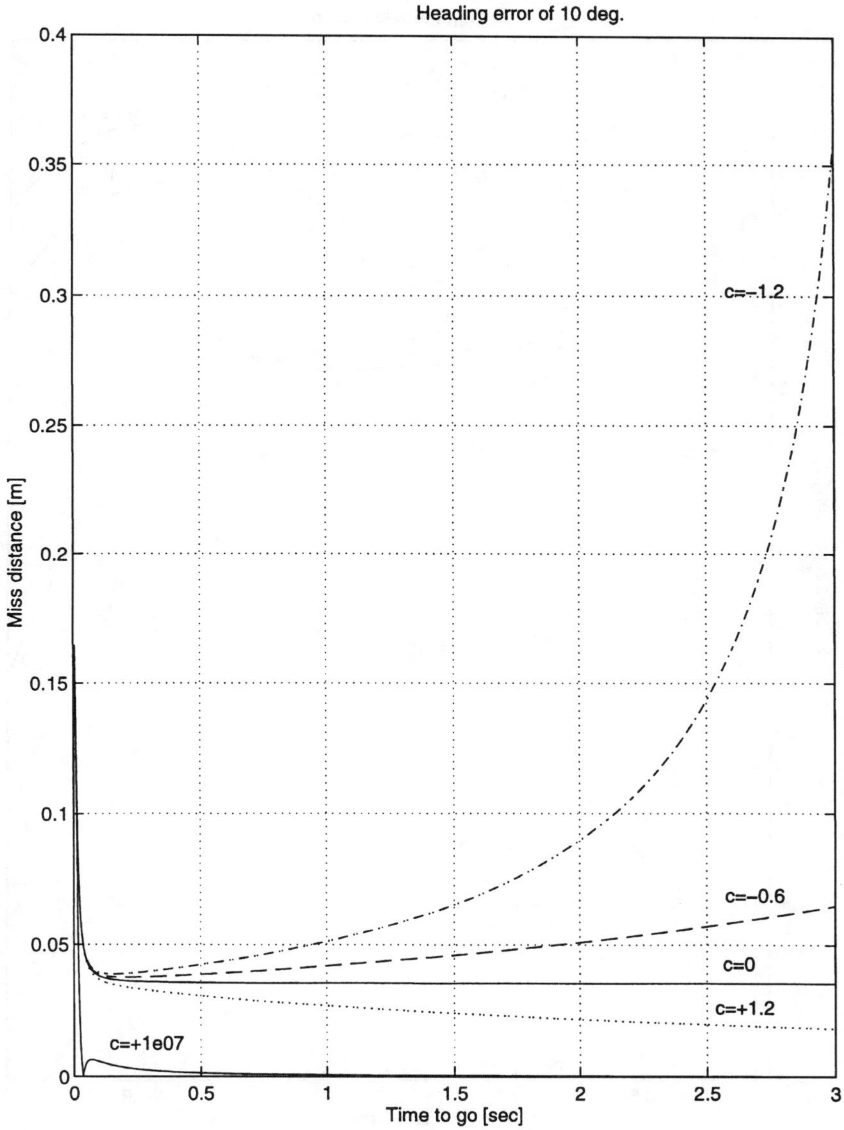

Fig. 3.11 Miss distance with trajectory shaping.

law is more moderate, leading, however, to larger miss distances. We consider the guidance laws that are given by Eqs. (3.48) for $1/b \in \{0,\ 0.1,\ 0.6,\ 1\}$, the pursuer's time constant is assumed to be 0.5 s, and the rest of the interception conflict parameters are as in Sec. VI.A. As in the previous section, we consider a couple of error sources, namely, 3 g of target (evader) acceleration and a 10-deg heading error. The simulation results are classified again according to the error sources. The results are illustrated by a comparison of the graphs of relative

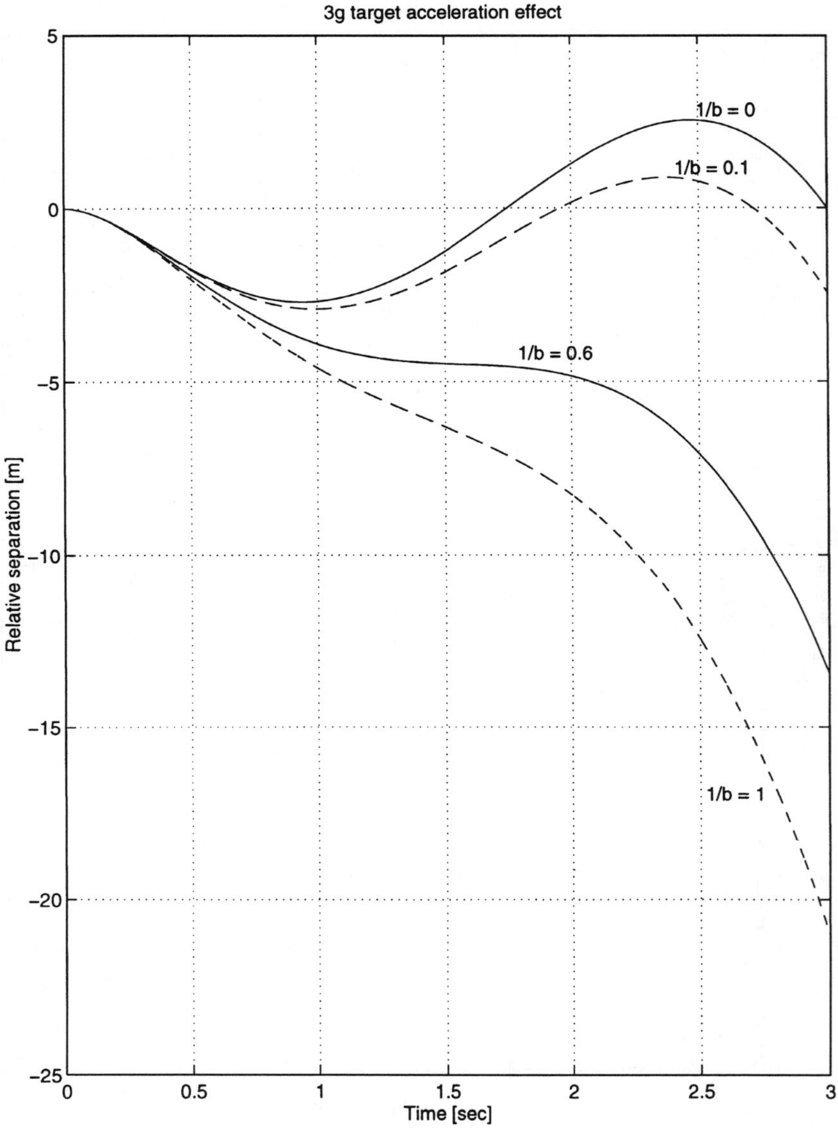

Fig. 3.12 Relative separation due to constant target maneuver—nonideal pursuer.

separation, relative angle (i.e., the quotient of the relative velocity and closing velocity in degrees), commanded pursuer acceleration, and $J_u(t) = \int_0^t u^2(\tau)\,d\tau$. A maximum commanded pursuer acceleration of 15 g is used in the simulations.

1. Step of 3 g *in Target Acceleration*

Notice from Fig. 3.12 that for $1/b = 0$ (i.e., infinite weight on miss distance) the miss distance is zero, whereas for larger $1/b$, the miss distances increase. For

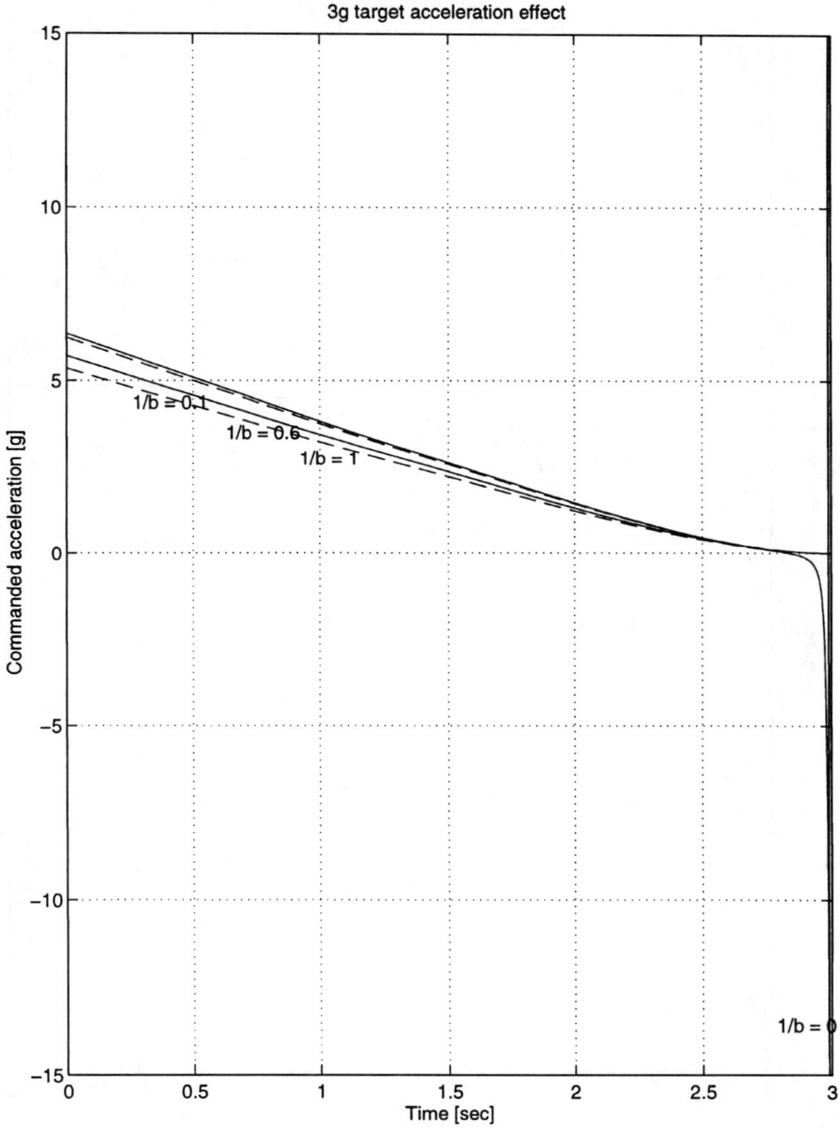

Fig. 3.13 Missile acceleration command due to constant target maneuver—nonideal pursuer.

$1/b = 1$, the miss distance is as large as about 20 m. However, lowering the miss distances by using small $1/b$ is not without cost. This is seen in Fig. 3.13, where the maximum required acceleration is 15 g (the limit we have imposed on the commands) for $1/b = 0$, whereas for $1/b = 0.1$ it is just about 7 g with a miss distance of 2 m. The normalized cost $j_u(t) = \int_0^t u^2(\tau)\,d\tau$ is depicted in Fig. 3.14 for all four cases, and again the effect of $1/b$ on the cost is as expected: the smaller $1/b$ (i.e., demand for small miss distance), the larger the cost. Thus it can

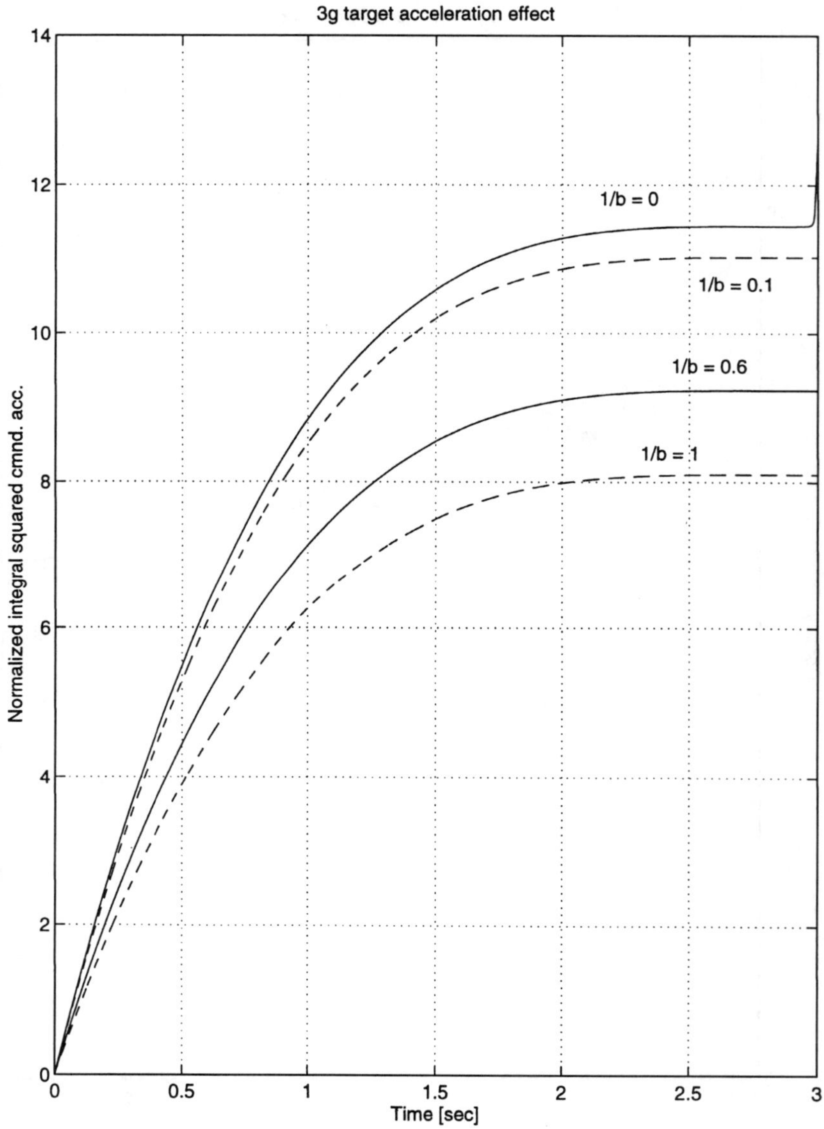

Fig. 3.14 Control effort due to constant target maneuver—nonideal pursuer.

be concluded that $1/b$ can be used as a design parameter that allows a trade-off involving miss distances and control efforts.

2. *Heading Error*

With an initial heading error only, the effect of $1/b$ is similar to its effect on target acceleration. Also here, $1/b$ seems to be an efficient design parameter that results in small miss distances at the cost of larger control effort when reduced. The results for this case are depicted in Figs. 3.15–3.17.

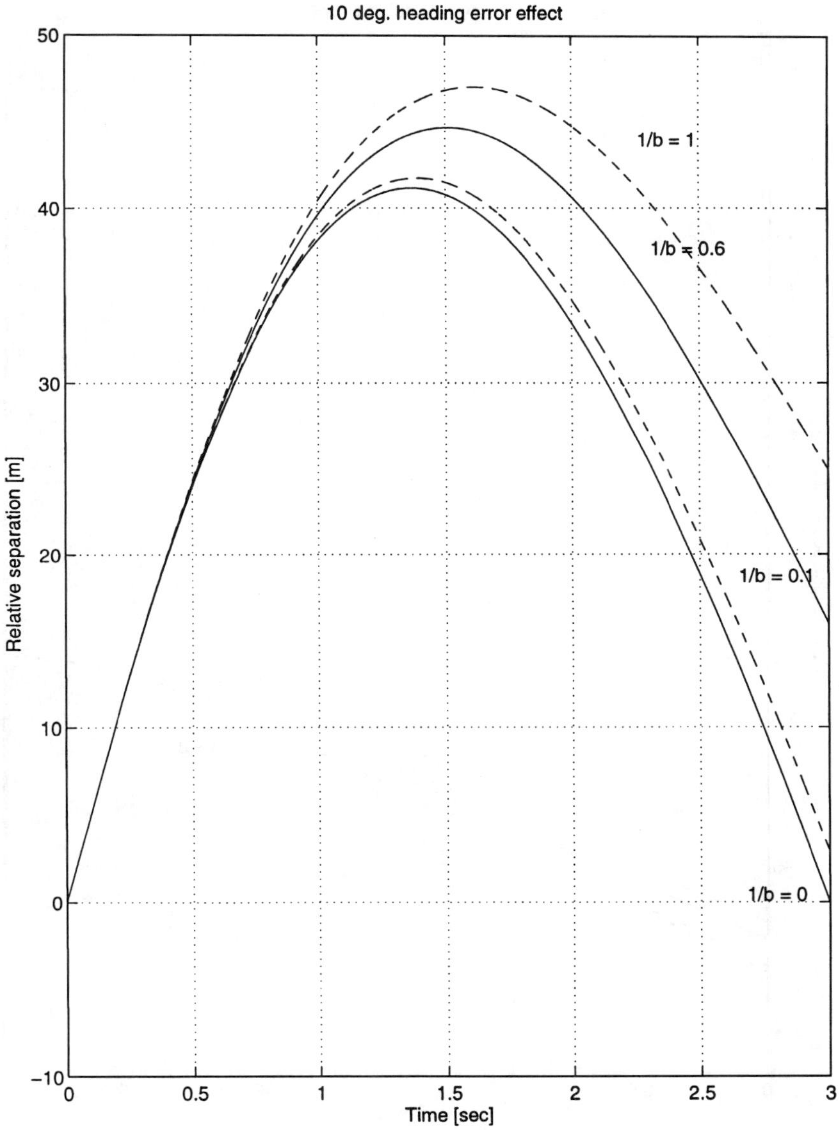

Fig. 3.15 Relative separation due to heading error—nonideal pursuer.

C. Nonideal Pursuer and Ideal Evader—The Effect of a Neglected Time Constant

The deleterious effects of guiding a nonideal pursuer having a significant time constant, in terms of the pursuit-evasion conflict duration, have lead to the development of the celebrated MEL.[2] If such a significant time constant is neglected, that is, a guidance law for an ideal pursuer is used, then large miss-distance values can result. For the sake of completeness we provide an example similar to that of

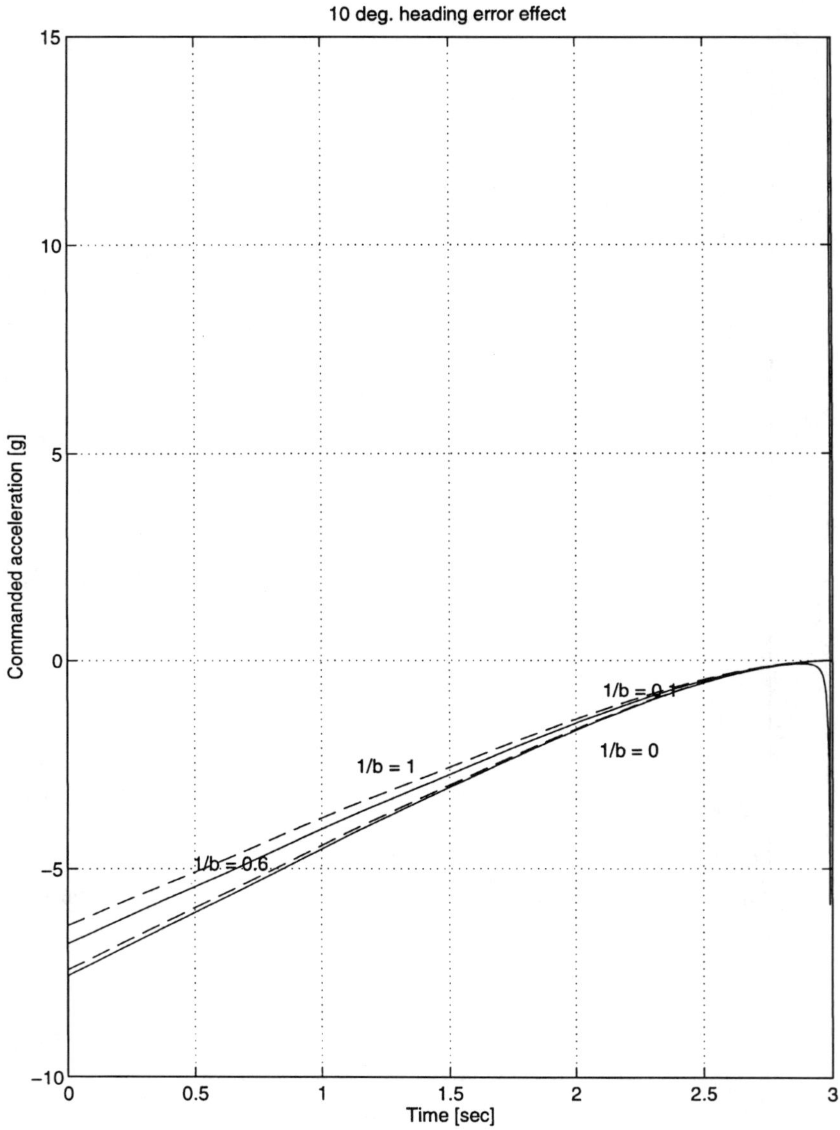

Fig. 3.16 Missile acceleration command due to heading error—nonideal pursuer.

Ref. 2 to illustrate the effect of a neglected time constant. We will compare the MEL to a couple of commonly used guidance laws, namely PN and APN (with a PN ratio of 3), in the derivation of which the time constant is neglected. The parameters of the conflict are as in Sec. VI.B. Notice that the time constant of 0.5 s is only 6 times smaller than the conflict duration of 3 s.

1. Step of 3 g *in Target Acceleration*

The results of forward simulations for a step of 3 *g* in target acceleration are given in Figs. 3.18–3.20. As might be expected, the PN shows large relative separation

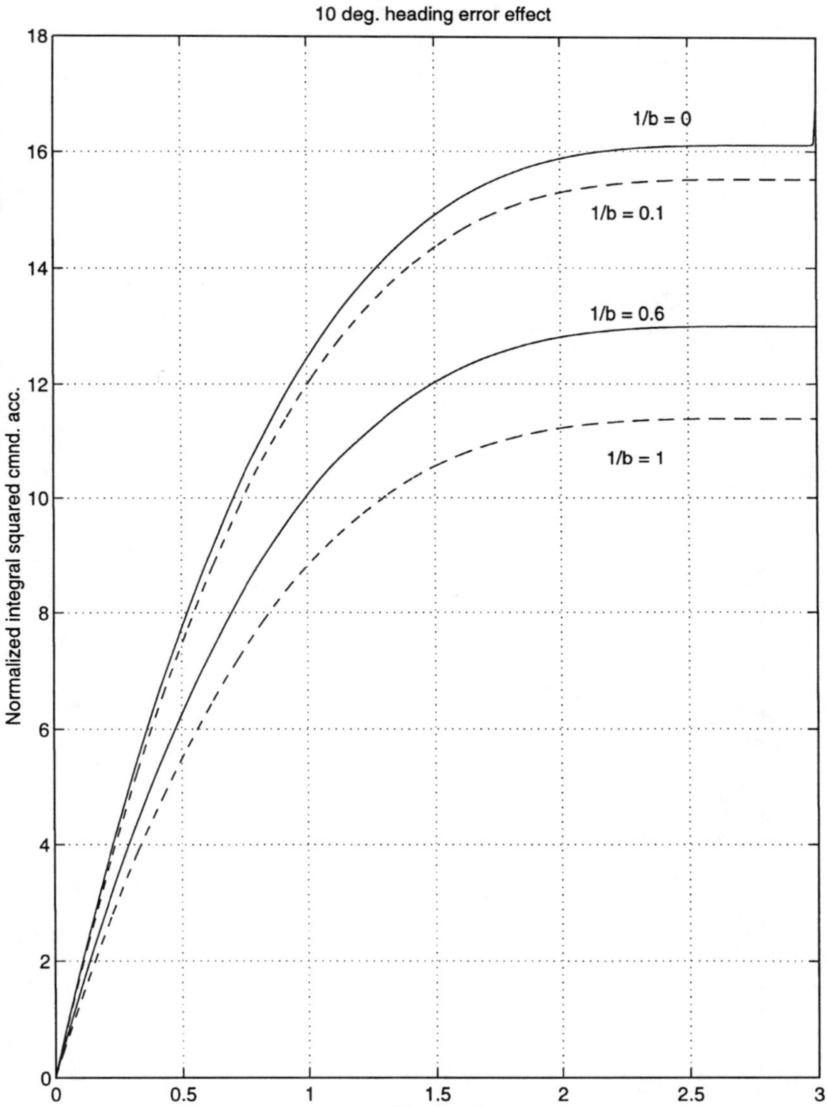

Fig. 3.17 Control effort due to heading error—nonideal pursuer.

during the conflict (see Fig. 3.18), while APN shows smaller separation due to the fact that it applies "feed forward" using the target acceleration measurement. Both PN and APN result, however, in nonzero miss distance (about 0.5 m and 0.7 m), while the MEL results in a near-zero miss distance. The most impressive advantage of the MEL is its efficient use of control effort. It does not require acceleration above about 6 g until about 0.1 s before hit, while both PN and APN use more than this in the last 50, and 10%, respectively, of the conflict (see Fig. 3.19). The results seem even more impressive in Fig. 3.20, where the integral-squared control signal is given. It is seen there that the control effort that is associated with the MEL is

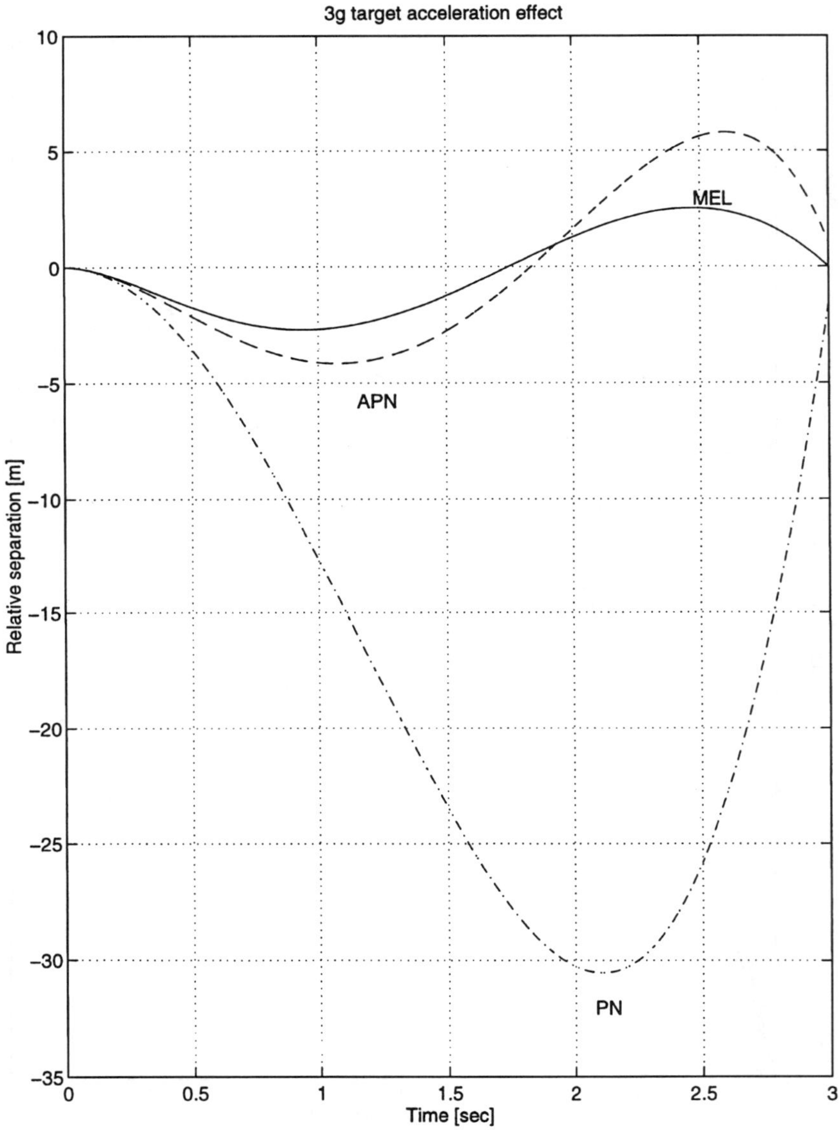

Fig. 3.18 The effect of neglected time-constant relative separation due to constant target maneuver.

considerably smaller than with PN and APN. In Fig. 3.21 the results of the adjoint simulation are depicted, where the miss distance is given as a function of the time-to-go at which the target applies its maneuver. The MEL is clearly superior to PN and APN. Since these guidance laws do not take into account the pursuer's time constant there are time-to-go's where it is more favorable for the target to apply maneuvers. Such time-to-go's are predicted by the peaks in Fig. 3.21. For PN there is a single peak that is observed at about 1 s to go (i.e., about 2 time constants),

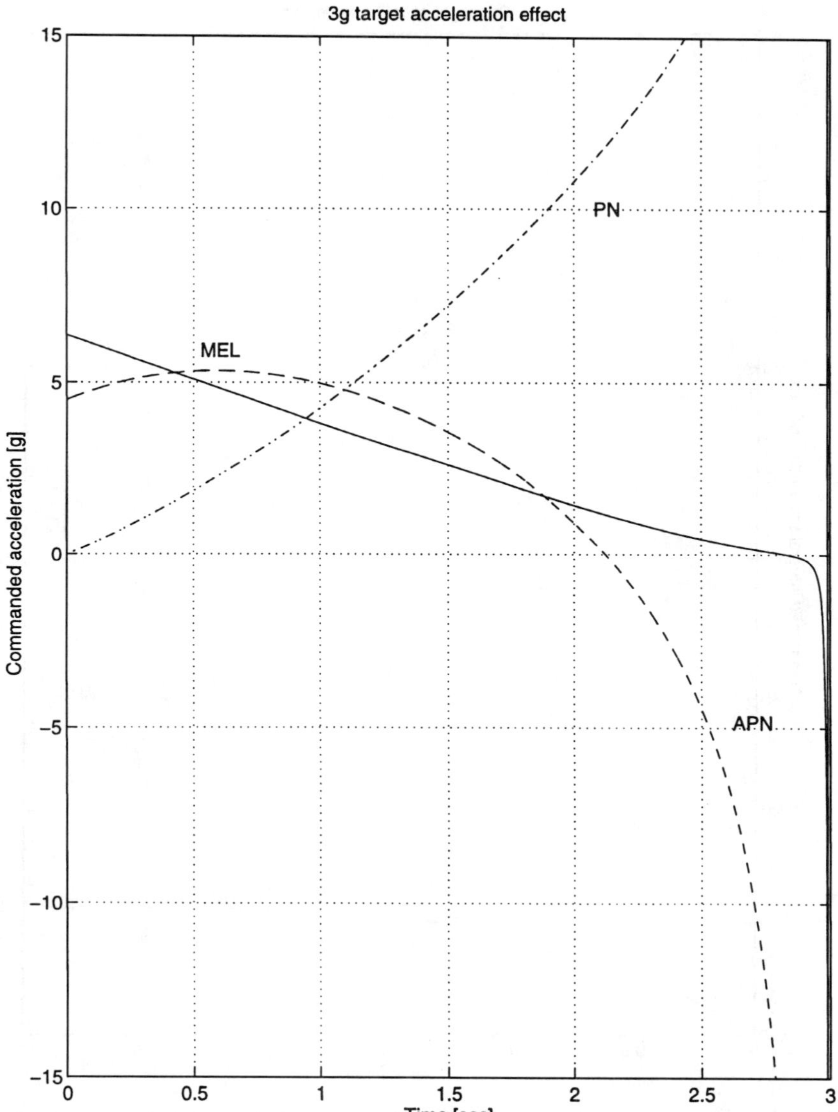

Fig. 3.19 The effect of neglected time-constant missile acceleration command due to constant target maneuver.

and for APN there are two peaks at about 0.5 s and 2 s to go. The result for PN can be qualitatively explained. If the maneuver is applied too late (i.e., small time-to-go) than the target does not move much until the end of the conflict and the miss distance is small. If, on the other hand, the maneuver is applied too soon (i.e., large time-to-go) then the pursuer has time to close the relative separation despite its time constant. The result with the APN where two peaks are observed is less obvious.

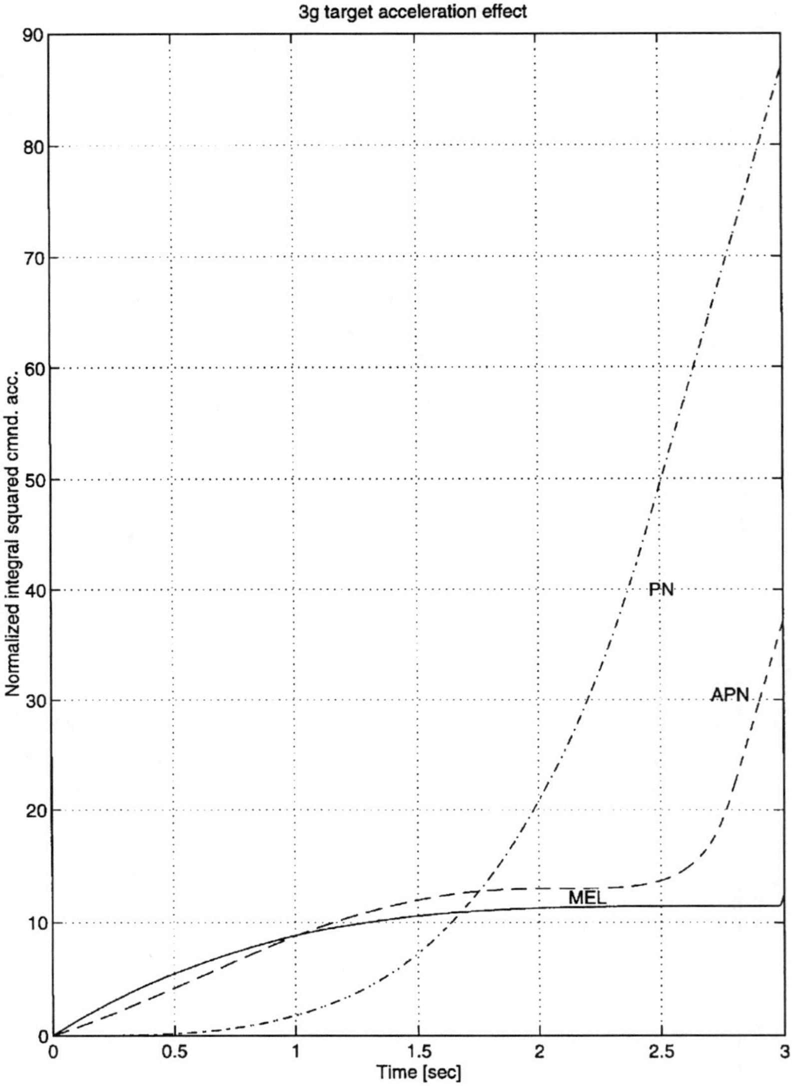

Fig. 3.20 The effect of neglected time-constant control effort due to constant target maneuver.

Remark: For PN with PN constants that are higher than 3, more than one peak is observed (see Ref. 2). It also is shown in Ref. 2 that optimal evasive maneuvers of the target, which are confined to be piecewise constant, are characterized by steps of alternating signs at each time-to-go, which corresponds to a peak in the miss distance graph of Fig. 3.21.[7] The rationale for this can be seen by noting that in Fig. 3.21 the absolute value of the miss distance is given; in fact, adjacent peaks have opposite signs.

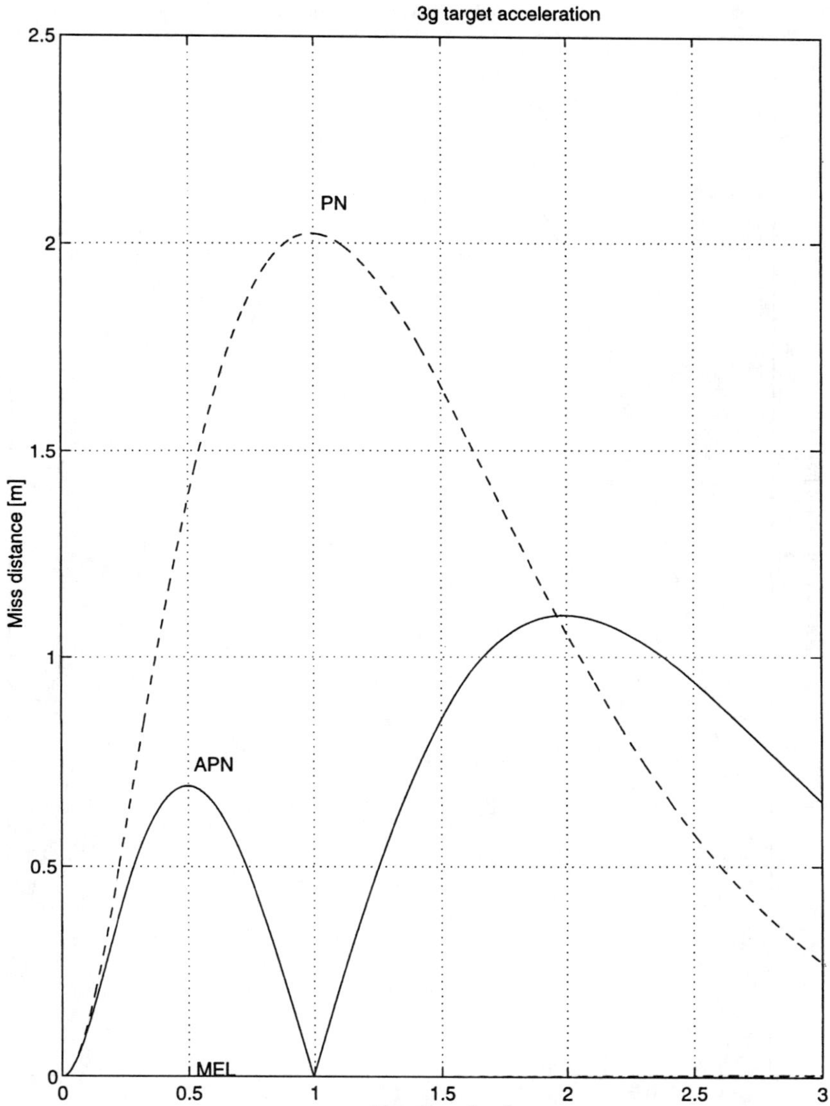

Fig. 3.21 The effect of neglected time-constant miss distance due to constant target maneuver.

2. *Heading Error*

In this case the PN and APN guidance laws coincide. In Fig. 3.22 we see that the MEL again results in a near-zero miss distance, whereas the PN/APN leads to a miss distance in the order of 1 m. The values of the relative separation throughout the conflict are of the same order with MEL and PN/APN. It is seen in Fig. 3.23 that the MEL applies smaller maneuvers than the PN/APN to get this smaller miss

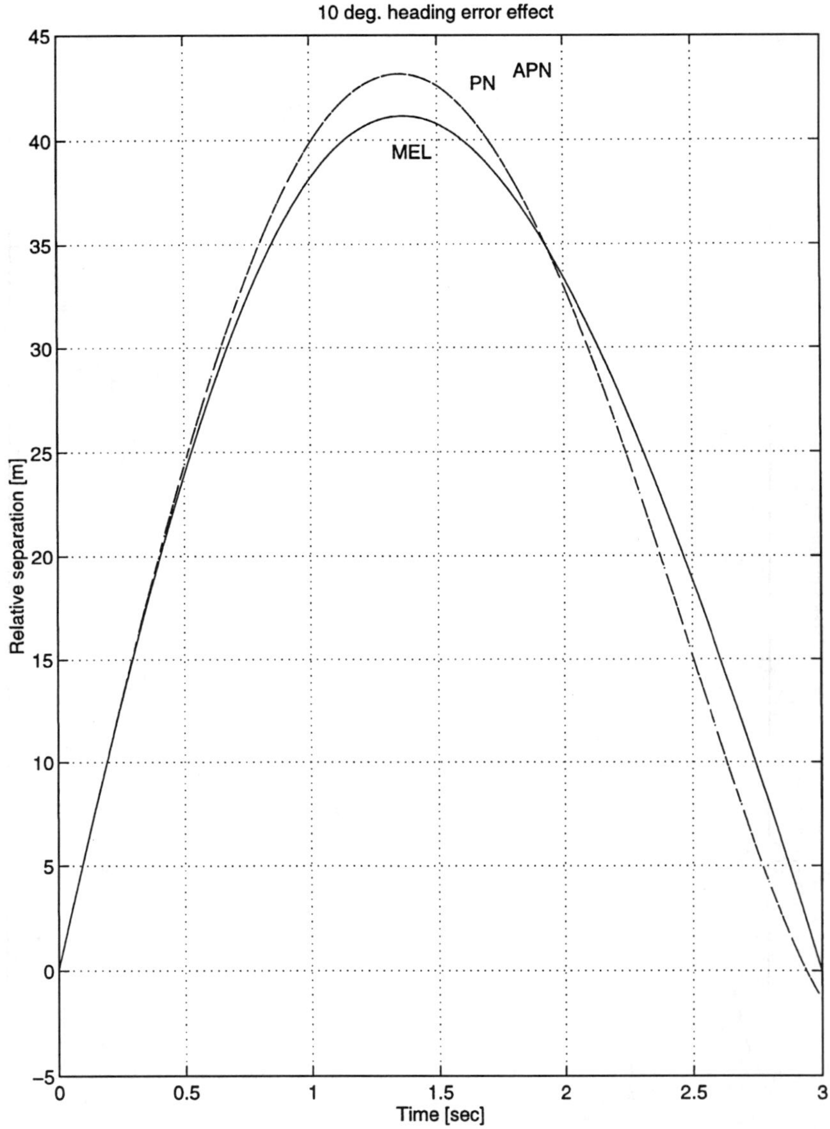

Fig. 3.22 The effect of neglected time-constant relative separation due to heading error.

distance; in Fig. 3.24 it is also observed that the overall control effort with MEL is smaller. In Fig. 3.25 the results of the adjoint simulation for heading error are depicted. (For commonality in form with the other adjoint results we also use time-to-go with the interpretation of total interception time.) It is seen that MEL provides very small miss distances for all conflict duration up to 3 s, while PN/APN are very ineffective with respect to the heading error with two large peaks of miss distance.

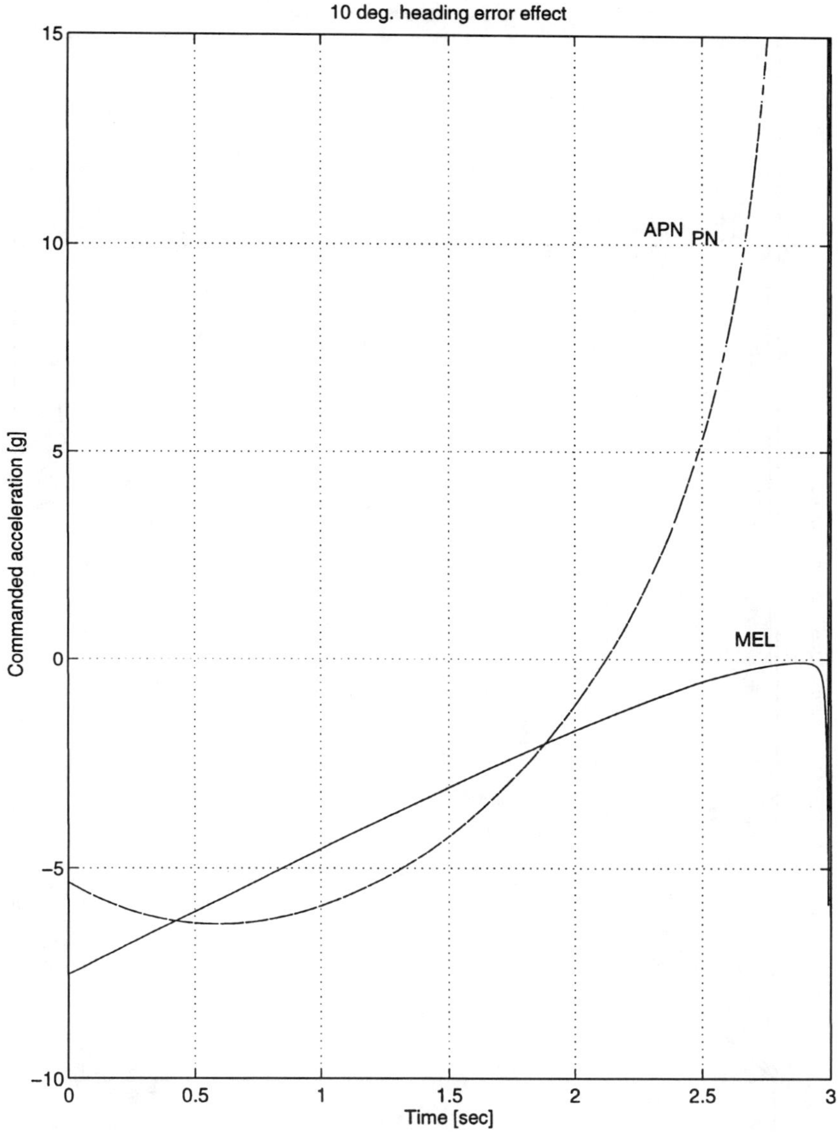

Fig. 3.23 The effect of neglected time-constant missile acceleration command due to heading error.

Remark: One should be careful about the implications of the impressive results of the MEL. These results were obtained based on the assumption that the time-to-go is exactly known to the pursuer. This is, however, not always the case. It was shown[8] that measurement errors in the time-to-go result in performance degradation. For large enough measurement errors PN/APN laws may become better than the MEL. We refer the reader to Chap. 5 in the book, where the issue of errors in the time-to-go is considered.

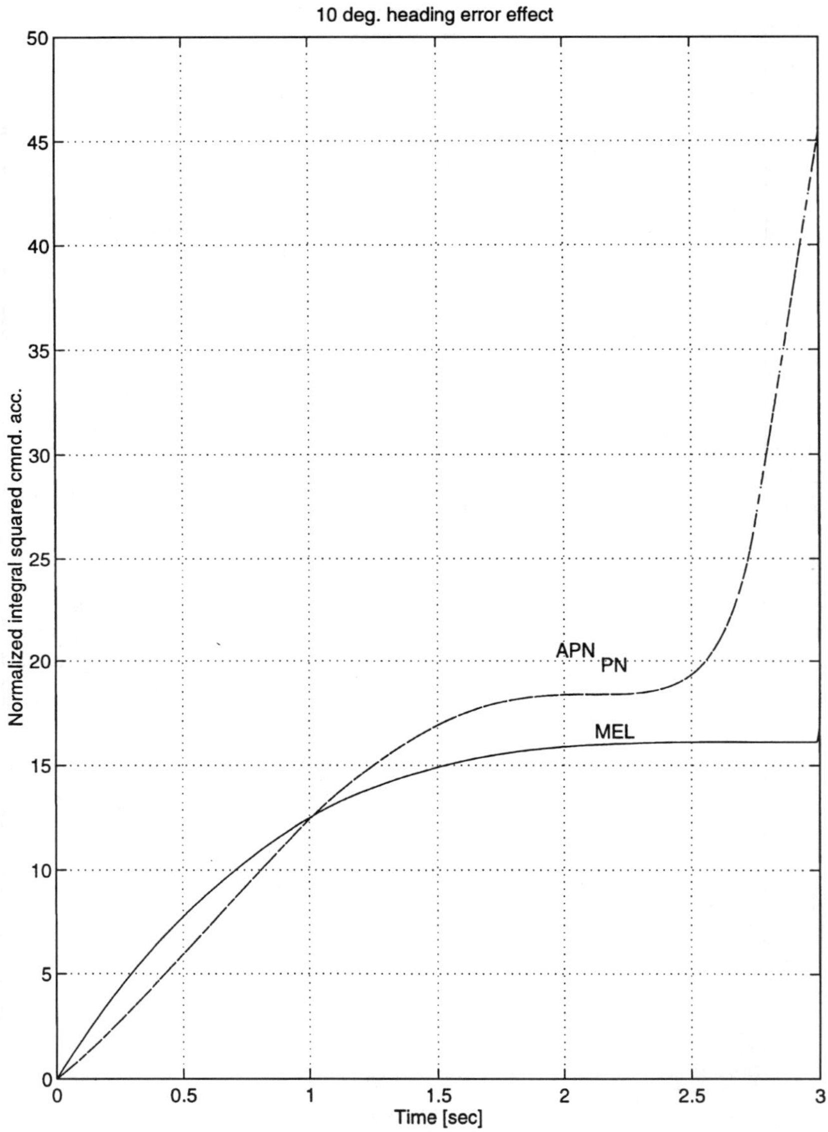

Fig. 3.24 The effect of neglected time-constant control effort due to heading error.

D. Nonideal Pursuer and Evader—The Effect of Random Maneuvers

In this section a numerical example is examined to illustrate the guidance law of Sec. IV.C, which takes into consideration the time constants of both the pursuer and the evader. As previously indicated, for $w_0 = 0$ the guidance law of Sec. IV.C was shown by Ref. 6 to be optimal for a first-order Markov process maneuver (i.e., when w is represented by a white noise process). In our formulation it is equivalent to the situation where, for design purposes, we assumed $w_0 = 0$, but in reality the

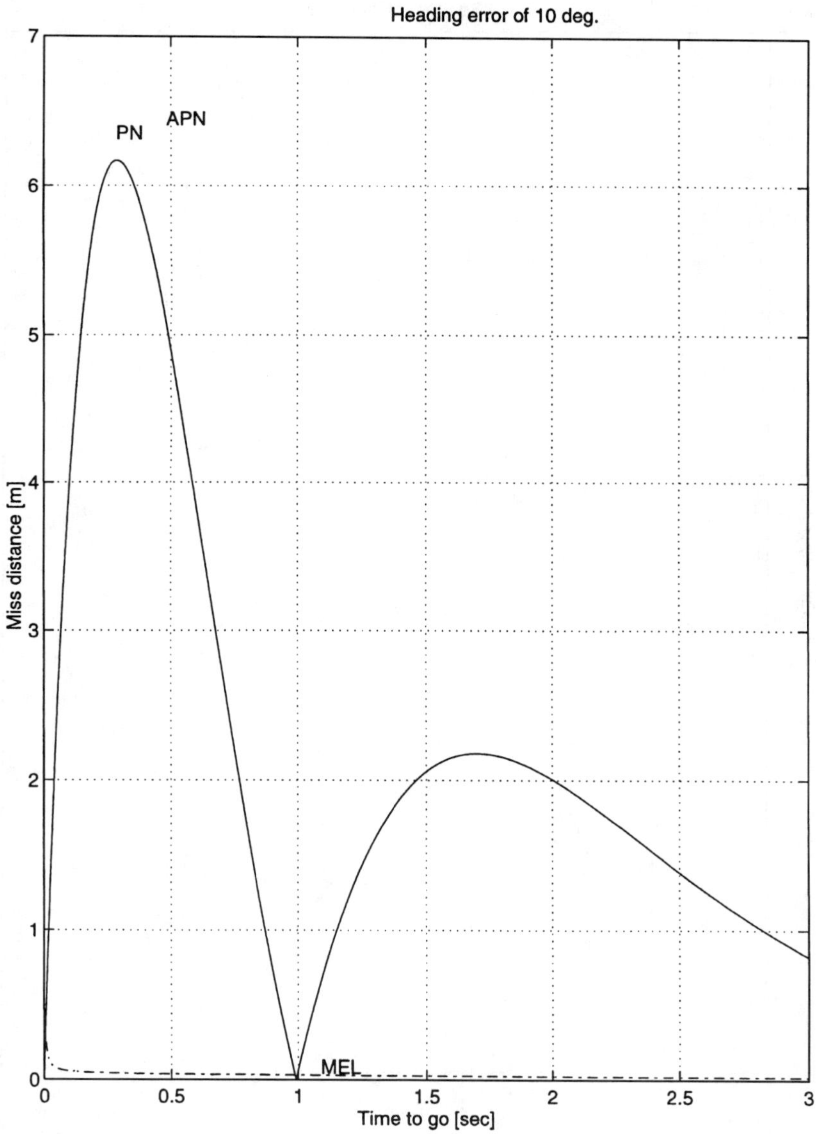

Fig. 3.25 The effect of neglected time-constant miss distance due to heading error.

evader's maneuvers are random. One way to cope with such random maneuvers is to assume that the evasive maneuver model is constant. Such an approach leads to the MEL.[2] The drawback of this approach for the case where the evader does in fact apply random evasive maneuvers can stem from the underlying assumptions of the MEL, especially the "weak" assumption that the evader's maneuvers are constant. Equivalently, this assumption means that the evader's time constant is of the same order of magnitude as the conflict duration. If this assumption is violated,

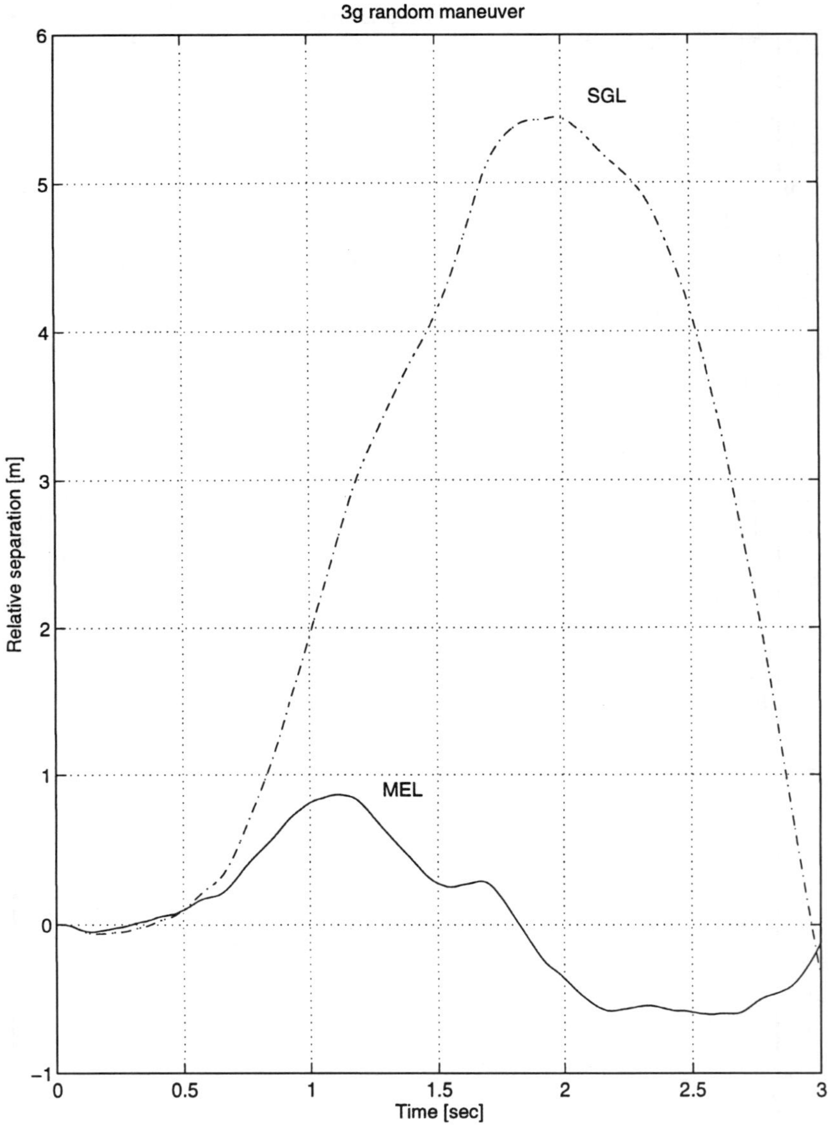

Fig. 3.26 Relative separation due to random target maneuver.

however, early evasive maneuvers should be overlooked by the pursuer. For this case we expect that the SGL guidance law of Sec. IV.C will be superior to the MEL. Therefore, our comparison will be of the two guidance laws: MEL and SGL.

The parameters of the conflict are that the interception conflict takes 3 s and it is assumed to be tail chase or head-on, with a closing velocity of 300 m/s. Again, we will use the linearized model of Eqs. (3.50) of the pursuit-evasion conflict, but take a 50 g maximum acceleration command for the pursuer. The pursuer's time

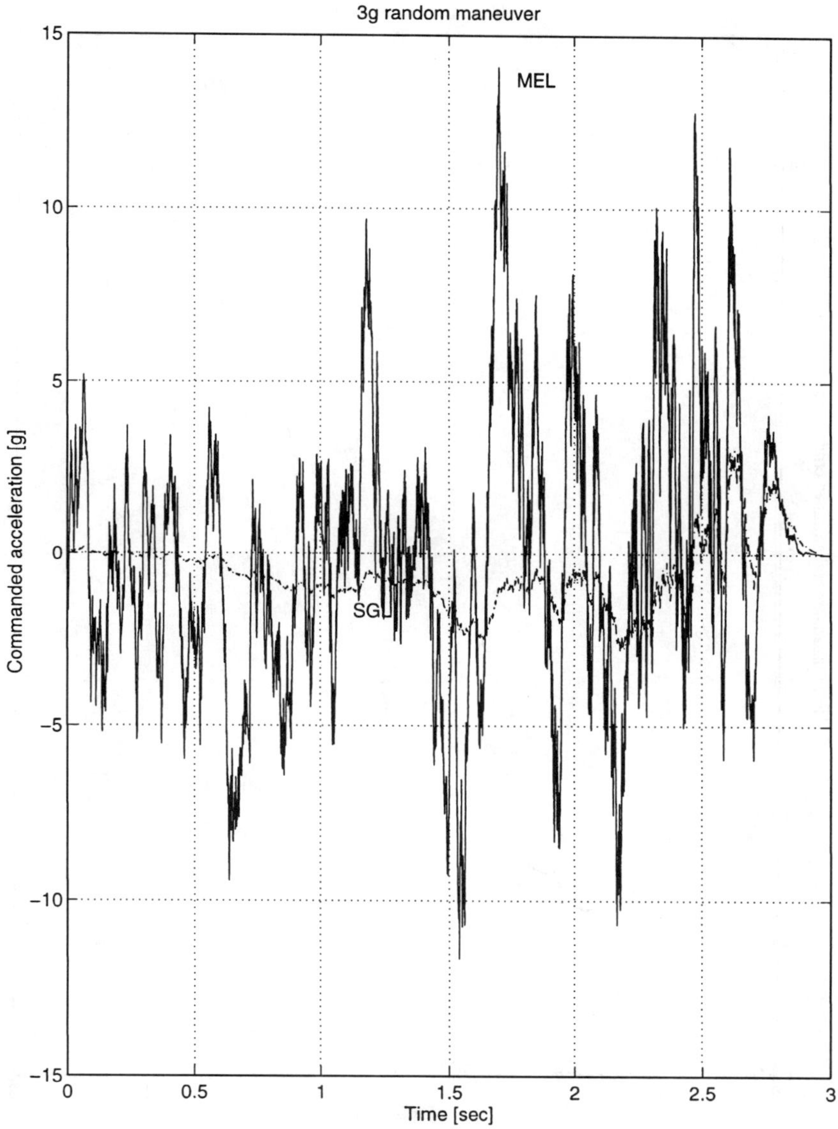

Fig. 3.27 Missile acceleration command due to random target maneuver.

constant is taken to be 0.5 s. For the target we assume a first-order Markov process with a time constant of 0.05 s and a standard deviation of 1.5 g.

The comparative simulation results are depicted in Figs. 3.26–3.28. In Fig. 3.26 the relative separation for both guidance laws is shown. The MEL shows considerably smaller values of relative separation than the SGL, but at the cost of much larger maneuvers (see Fig. 3.27). The reason for this is obvious: When the MEL "detects" a target acceleration, it assumes that it will stay constant throughout the interception conflict. Therefore, its strategy is immediately to close the relative

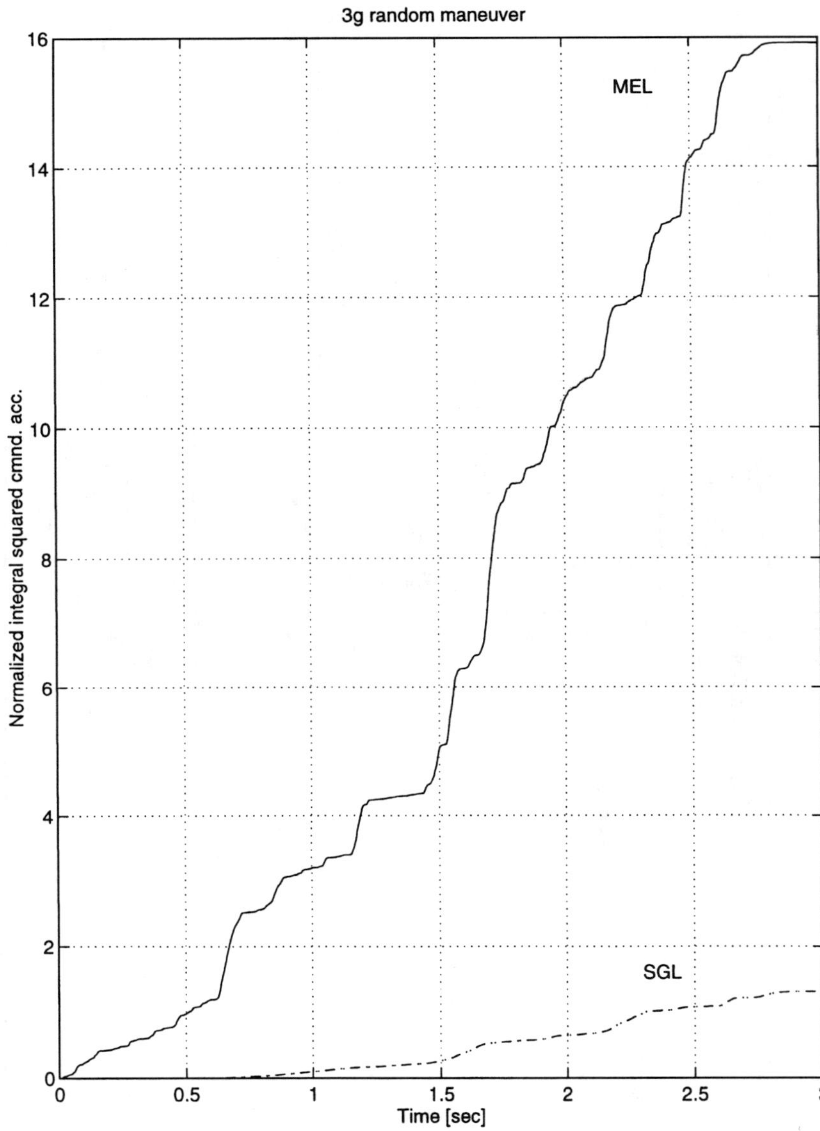

Fig. 3.28 Control effort due to random target maneuver.

position so as to stay "on" the target. However, the target continues to alter its acceleration very quickly, thus requiring the MEL to demand large pursuer maneuvers. On the other hand, the SGL "knows" that the evader's accelerations will not remain constant at any time and saves its energy for the conflict end time. Overall, the MEL puts in much more control effort than the SGL (see Fig. 3.28). It should be noted that the superiority of the SGL is in the energy it uses and not in the miss-distance values that are comparable to those of the MEL. In Fig. 3.29 the results of an adjoint simulation (which does not take the pursuer acceleration

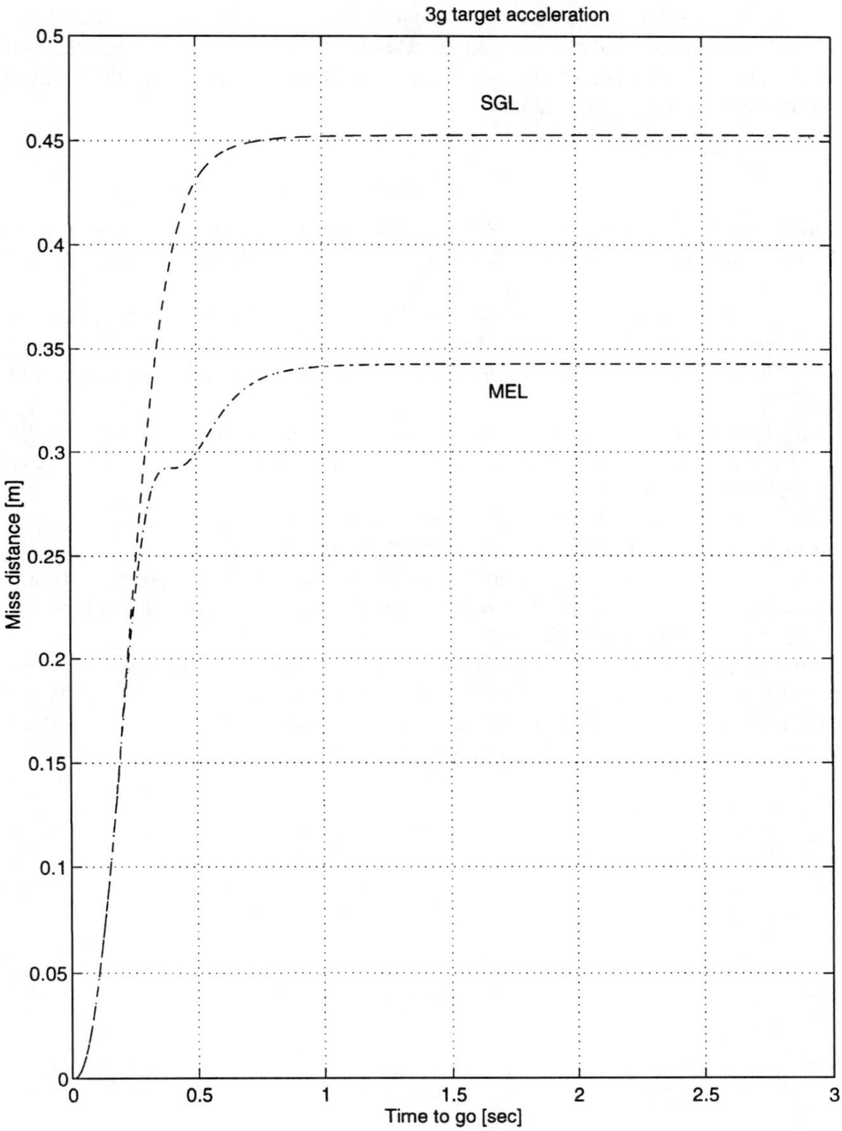

Fig. 3.29 Miss distance due to random target maneuver.

limit into account) are given. The rms miss-distance values of the MEL and SGL are comparable with MEL being slightly better.

VII. Concluding Remarks

In this chapter a systematic derivation of one-sided optimal control based guidance laws was developed. Along the way, many well-known guidance schemes, such as PN, APN, and MEL were revisited. Methods that weight the terminal velocity and acceleration also were presented along with the ensuing possibilities of trajectory shaping. The results can be extended to other cases (e.g., constant jerk rather than constant acceleration as the evasive maneuver) along the same lines to obtain more closed-form guidance laws.

References

[1]Anderson, G. M., "Comparison of Optimal Control and Differential Game Intercept Missile Guidance Laws," *Journal of Guidance and Control*, Vol. 4 , No. 2, 1981, pp. 109–115.

[2]Zarchan, P., *Tactical and Strategic Missile Guidance,* Vol. 124, Progress in Astronautics and Aeronautics, AIAA, Washington, DC, 1990, Chap. 7.

[3]Bryson, A. E., and Ho, Y. C., *Applied Optimal Control,* Hemisphere, New York, 1975, Chap. 5.

[4]Gill, A., and Sivan, R., "Optimal Control of Linear Systems with Quadratic Costs Which Are Not Necessarily Positive Definite," *IEEE Transactions on Auto. Cont.,* Vol. AC-14, No. 1, 1969, pp. 83–88.

[5]Boyce, W. E., and DiPrima, R. C., *Elementary Differential Equations and Boundary Value Problems,* 3rd ed., Wiley, New York, 1977, Chap. 4.

[6]Loveren, N., and Tomic, M., "Analytic Solution of the Riccati Equation for the Homing Missile Linear-Quadratic Control Problem," *Journal of Guidance, Control, and Dynamics,* Vol. 17, No. 3, 1994, pp. 619–621.

[7]Shinar, J., and Steinberg, D., "Analysis of Optimal Evasive Maneuvers Based on a Linearized Two-Dimensional Model," *Journal of Aircraft,* Vol. 14, No. 4, 1977, pp. 795–802.

[8]Nesline, W. N., and Zarchan, P., "A New Look at Classical vs Modern Homing Missile Guidance," *Journal of Guidance and Control,* Vol. 4, No. 1, 1981, pp. 78–84.

Chapter 4

Optimal Guidance Based on Linear-Quadratic Differential Games

I. Introduction

REFERENCE 1 compared differential games with one-sided optimal methods and advocated the use of differential game guidance methods, particularly against maneuvering targets. In a well-known differential game (originally solved by Baron[2]) the target was allowed to maneuver to maximize the miss distance and the missile's control effort while minimizing its own control effort, and simultaneously the pursuer tried to minimize the same cost function, thus forming a "zero sum" game.[3] The strategy obtained in this classical work will be referred to as the game-theoretic optimal intercept.

This chapter proposes some extensions to this strategy as well as some more general results. As before, we begin by assuming an ideal pursuer and an ideal evader according to our previous definition. The differential game approach is used where the terminal lateral velocity is introduced into the cost in addition to the adversaries' control efforts and the miss distance. The case of zero terminal conditions will naturally be called the game-theoretic optimal rendezvous, and it will be shown that for this case the optimal strategies as well as the resulting optimal trajectories can be given in closed form. The assumption of the ideal pursuer will then be removed, and the problem will be reformulated and solved analytically. Following this, the assumption of the ideal evader will also be removed and the problem will be solved again. We will finally consider cases where the target's maneuver is composed of two parts: one constant ("dumb" maneuver) and known to the pursuer, and the other optimal with respect to the preceding cost ("smart" maneuver) and unknown to the pursuer.

II. Ideal Pursuer Case

A. General Solution

We use the problem formulation given previously, that is,

$$\dot{x} = Ax + Bu + Dw$$
$$A = \begin{bmatrix} 0 & 1 \\ 0 & 0 \end{bmatrix} \qquad B = \begin{bmatrix} 0 \\ 1 \end{bmatrix} \qquad D = \begin{bmatrix} 0 \\ 1 \end{bmatrix} \tag{4.1}$$

In this chapter, however, we allow the evader to perform an optimal evasive maneuver. The optimal minimax problem is, therefore, as follows:

$$\min_u \max_w J = \frac{b}{2}x_1^2(t_f) + \frac{c}{2}x_2^2(t_f) + \frac{1}{2}\int_{t_0}^{t_f} [u^2(t) - \gamma^2 w^2(t)]\,\mathrm{d}t \tag{4.2}$$

where $b > 0$ (the pursuer aims to minimize the miss and the evader to maximize it) and $\gamma > 1$.

The latter condition implies that the evader is less maneuverable than the pursuer, hence the penalty on its control effort is higher. We, therefore, have a zero-sum game.[3] As before, in game-theoretic perfect intercepts (with or without velocity constraints) we will assume that $b \to \infty$, and finally for the game-theoretic optimal rendezvous we will assume that $b \to \infty$ and $c \to \infty$, forcing both $x_1(t_f) = 0$ and $x_2(t_f) = 0$.

The adjoint equations remain

$$\dot{\lambda}_1 = 0 \qquad \dot{\lambda}_2 = -\lambda_1 \tag{4.3}$$

but from the minmax operation we get

$$u = -\lambda_2 \qquad w = \gamma^{-2}\lambda_2 \tag{4.4}$$

(see Chapter 2 for details). Similar to Sec. III.A of Chapter 3, we employ the terminal conditions to infer that

$$\begin{aligned} \lambda_1 &= bx_1(t_f) \\ \lambda_2 &= bx_1(t_f)(t_f - t) + cx_2(t_f) \end{aligned} \tag{4.5}$$

Substituting $u = -[bx_1(t_f)(t_f - t) + cx_2(t_f)]$ and $w = \gamma^{-2}[bx_1(t_f)(t_f - t) + cx_2(t_f)]$ into Eqs. (4.1), and then integrating it from t_0 to $t = t_f$, we get implicit expressions for $x_1(t_f)$ and $x_2(t_f)$. Now we can solve the resulting linear algebraic equations to get explicit expressions for the unknown terminal values $x_1(t_f)$ and $x_2(t_f)$. Having found them, we get an expression for $u(t)$ [and also for $w(t)$], and in particular an expression for $u(t_0)$ that is linear in $x_1(t_0)$ and $x_2(t_0)$. Since t_0 is arbitrary we may eliminate the subscript 0 to obtain the following linear feedback form:

$$u = -(g_1 x_1 + g_2 x_2)$$

$$g_1 = \frac{b\gamma^2\tau\left[\frac{1}{2}\tau c(\gamma^2 - 1) + \gamma^2\right]}{\tau^4(cb/12)(\gamma^2 - 1)^2 + [\tau c + \tau^3(b/3)]\gamma^2(\gamma^2 - 1) + \gamma^4} \tag{4.6}$$

$$g_2 = \frac{\frac{1}{3}\gamma^2[\tau^3 cb(\gamma^2 - 1) + 3(\tau^2 b + c)\gamma^2]}{\tau^4(cb/12)(\gamma^2 - 1)^2 + [\tau c + \tau^3(b/3)]\gamma^2(\gamma^2 - 1) + \gamma^4}$$

where as previously stated $\tau = t_f - t$.

Listing 4.1 presents a MAPLE routine to carry out the derivation of Eqs. (4.2–4.6).

Listing 4.1 Gains for ideal pursuer and evader

```
> # listing 4.1
> # l1 is constant and equals b*x1f
> #the following is the remaining costate equation for l1, namely:dl2/dt=−l1
> #
> sys:=diff(l2(t),t)+b*x1f;
> dsolve({sys,l2(tf)=c*x2f},l2(t));
> assign(");
> simplify(");
> #
> # state equation for x2, namely: dx2/dt=u + w=−l2+l2/beta^2
> #
> sys1:=diff(x2(t),t)+(1−beta^(−2))*l2(t);
> dsolve({sys1,x2(0)=v0},x2(t));
> assign(");
> simplify(");
> #
> # state equation for x1, namely: dx1/dt=x2
> #
> sys3:=diff(x1(t),t)−x2(t);
> dsolve({sys3,x1(0)=x0},x1(t));
> assign(");
> #
> # evaluation and solving for x1f, x2f
> #
> x1(t);
> y1:=subs(t=tf,");
> x2(t);
> y2:=subs(t=tf,");
> sys5:={x1f−y1,x2f−y2};
> solve(",{x1f,x2f});
> assign(");
> simplify(");
> #
> # evaluating the control u=−l2 at t=0
> #
> u:=−eval(l2(t));
> u0:=subs(t=0,");
> simplify(");
> #
> # limit values for intercept and for perfect rendezvous
> #
> limit(",b=infinity);
> limit(",c=infinity);
```

Remarks:

1) The Riccati equation for this problem is the following[4]:

$$-\dot{P} = PA + A^T P - PBB^T P + \gamma^{-2} PDD^T P \tag{4.7}$$

with the terminal conditions

$$P(t_f) = \begin{bmatrix} b & 0 \\ 0 & c \end{bmatrix}$$

In terms of P the optimal strategies are then of the form

$$\begin{aligned} u &= -B^T P x \equiv -g_1 x_1 - g_2 x_2 \\ w &= D^T \gamma^{-2} P x = \gamma^{-2}(g_1 x_1 + g_2 x_2) \end{aligned} \tag{4.8}$$

By solving the linear equation for $S \equiv P^{-1}(\dot{S} = AS + SA^T - BB^T + \gamma^{-2} DD^T)$ with the appropriate terminal conditions, we can obtain the feedback gains in Eqs. (4.6). Listing 4.2 derives these values analytically using MAPLE.

2) Notice, again, that even though $\gamma > 1$, a bounded solution here is not guaranteed for $c < 0$.

3) By exactly the same arguments of Sec. III.A in Chapter 3, the solution does not contain a conjugate point if and only if the gains are finite. Notice that this provides a sufficient condition for optimality (although in this case it is not a necessary condition; see Chapter 2).

4) Also note that Eqs. (3.10) can be obtained from Eqs. (4.4) simply by letting $\gamma \to \infty$.

B. Case Without Velocity Constraints

Substituting $c = 0$ in Eqs. (4.6) (i.e., no velocity constraints), we obtain

$$\begin{aligned} g_1 &= \frac{\tau}{(\tau^3/3)(1 - \gamma^{-2}) + (1/b)} \\ g_2 &= \frac{\tau^2}{(\tau^3/3)(1 - \gamma^{-2}) + (1/b)} \end{aligned} \tag{4.9}$$

In this case no conjugate point exists for our choice of γ and b and the gains are bounded. Having chosen $\gamma < 1$ might yield a conjugate point for some t_f (see also Ref. 5). By letting $b \to \infty$ (imposing zero miss), we get

$$\begin{aligned} g_1 &= \frac{3}{\tau^2(1 - \gamma^{-2})} \\ g_2 &= \frac{3}{\tau(1 - \gamma^{-2})} \end{aligned} \tag{4.10}$$

Thus $\{g_1, g_2\}$ are the feedback gains for the game-theoretic optimal intercept, which can be viewed as a generalization of the PN case. Notice that the equivalent gain is higher than the PN gain as demonstrated by the following representation

Listing 4.2 Solving Lyapunov equations—ideal case

```
> # listing 4.2
> with(linalg);
> #
> # state transition and control matrices
> #
> A:=array([[0,1],[0,0]]);
> B:=array(1..2,1..1,[[0],[1]]);
> #
> # constructing the Lyapunov equation for S=P^(−1), namely:
> # dS/dt=AS+S trans(A) − (1−beta^(−2))*B trans(B)
> #
> S(t):=array(1..2,1..2,[[S1(t),S2(t)],[S2(t),S3(t)]]);
> temp:=evalm((S(t)&*transpose(A))+(A)&*S(t));
> SD:=evalm((transpose(temp)−(1−beta^(−2))*(B&*transpose(B))));
> convert(SD,array);
> #
> # solving the Lyapunov equation
> #
> system:=diff(S1(t),t)=SD[1,1],diff(S2(t),t)=SD[1,2],diff(S3(t),t)=SD[2,2];
> system1:=S3(Tf)=1/c,S2(Tf)=0,S1(Tf)=1/b;
> dsolve({system,system1},{S1(t),S2(t),S3(t)});
> assign(");
> simplify(");
> # inverting S to find P and calculating the feedback gain trans(B) P
> #
> S(t):=array(1..2,1..2,[[S1(t),S2(t)],[S2(t),S3(t)]]);
> P(t):=inverse(S(t));
> G:=evalm(transpose(B)&*P(t));
> G1:=subs(c=0,");
```

[see Eqs. (3.14)]:

$$u = \frac{-N' V_c \dot{\lambda}}{(1-\gamma^{-2})} \qquad N' = 3$$

where optimal PN is obtained by taking $\gamma \to \infty$.

This result corresponds well with the common practice of using higher PN constants against maneuvering targets. [The smaller γ is, the more maneuverable the target is, and consequently, $N'/(1-\gamma^{-2})$ gets larger.]

The corresponding equations of motion, assuming both players are playing optimally, become

$$\begin{aligned} \dot{x}_1 &= x_2 \\ \dot{x}_2 &= u + w = -(1-\gamma^{-2})(g_1 x_1 + g_2 x_2) = \frac{-3}{\tau^2} x_1 + \frac{-3}{\tau} x_2 \end{aligned} \tag{4.11}$$

Notice the interesting result that the governing equations are identical to those of the one-sided optimal intercept with a nonmaneuvering target and are independent of γ. Equations (3.16) and (3.17) are, therefore, still valid and the trajectories satisfy

$$
\begin{aligned}
x_1 &= c_1\tau + c_2\tau^2 \\
x_2 &= -c_1 - 2c_2\tau \\
c_1 &= \frac{2}{t'_f}x_1(t_0) + x_2(t_0) \\
c_2 &= \frac{-1}{t'^2_f}x_1(t_0) - \frac{1}{t'_f}x_2(t_0)
\end{aligned}
\tag{4.12}
$$

where $t'_f = t_f - t_0$.

C. Perfect Interception with Terminal Velocity Constraints

Assume now that $c \neq 0$, and consequently the terminal relative velocity is of interest. Letting $b \to \infty$ in Eqs. (4.6) [i.e., $x_1(t_f) = 0$], we get

$$
\begin{aligned}
g_1 &= \frac{6\tau + [12/c(1-\gamma^{-2})]}{\tau^3(1-\gamma^{-2}) + (4\tau^2/c)} \\
g_2 &= \frac{4\tau + [12/c(1-\gamma^{-2})]}{\tau^2(1-\gamma^{-2}) + (4\tau/c)}
\end{aligned}
\tag{4.13}
$$

For the case $c > 0$, the solution exists for all $\tau \in (0, t_f)$. However, for $c < 0$ the solution exists only if

$$\tau < -\frac{4}{c(1-\gamma^{-2})}$$

Since the maximum value on the left-hand side is t_f, we obtain the condition

$$t_f < -\frac{4}{c(1-\gamma^{-2})}$$

Based on the preceding discussion, this condition becomes (as in the one-sided optimal control problem) the no-conjugate-point condition.

The perfect game-theoretic rendezvous is obtained directly from Eqs. (4.6) by letting $b \to \infty$ and $c \to \infty$. We obtain the following feedback gains

$$
\begin{aligned}
g_1 &= \frac{6}{\tau^2(1-\gamma^{-2})} \\
g_2 &= \frac{4}{\tau(1-\gamma^{-2})}
\end{aligned}
\tag{4.14}
$$

This very simple form constitutes a natural extension of Eqs. (3.12) and (4.10).

The corresponding equations of motion, assuming both players are playing optimally, become

$$\dot{x}_1 = x_2$$
$$\dot{x}_2 = u + w = -(1 - \gamma^{-2})(g_1 x_1 + g_2 x_2) = \frac{-4}{\tau^2} x_1 + \frac{-6}{\tau} x_2 \tag{4.15}$$

Thus, similarily to the game-theoretic optimal intercept case, the trajectories are γ-independent and can be expressed by Eqs. (3.22–3.23), namely,

$$\begin{aligned} x_1 &= c_1 \tau^2 + c_2 \tau^3 \\ x_2 &= -2c_1 \tau - 3c_2 \tau^2 \\ c_1 &= \frac{3}{t_f'^2} x_1(t_0) + \frac{1}{t_f'} x_2(t_0) \\ c_2 &= \frac{-2}{t_f'^3} x_1(t_0) + \frac{1}{t_f'^2} x_2(t_0) \end{aligned} \tag{4.16}$$

where $t_f' = t_f - t_0$.

III. Nonideal Pursuer Case

A. Ideal Target

Assuming, as before, a first-order lag for the pursuer's dynamics (with no such lag for the evader's dynamics), we get the following equations in state space:

$$\dot{x} = Ax + Bu + Dw$$

$$A = \begin{bmatrix} 0 & 1 & 0 \\ 0 & 0 & 1 \\ 0 & 0 & -\frac{1}{T} \end{bmatrix} \qquad B = \begin{bmatrix} 0 \\ 0 \\ \frac{1}{T} \end{bmatrix} \qquad D = \begin{bmatrix} 0 \\ 1 \\ 0 \end{bmatrix} \tag{4.17}$$

where the control u is now the commanded normal acceleration of the pursuer and where T is its time constant. We allow an evasive maneuver w, but, for simplicity, we do not consider any additional terminal weightings other than the miss distance. The problem, therefore, becomes

$$\min_u \max_w J = \frac{b}{2} x_1^2(t_f) + \frac{1}{2} \int_{t_0}^{t_f} [u^2(t) - \gamma^2 w^2(t)]\, dt$$

The adjoint equations and corresponding terminal conditions are as follows:

$$\dot{\lambda}_1 = 0 \qquad \dot{\lambda}_2 = -\lambda_1 \qquad \dot{\lambda}_3 = -\lambda_2 + \frac{\lambda_3}{T}$$

$$u = -\frac{1}{T}\lambda_3 \qquad w = \gamma^{-2}\lambda_2 \tag{4.18}$$

$$\lambda_1(t_f) = b x_1(t_f) \qquad \lambda_2(t_f) = \lambda_3(t_f) = 0$$

The latter relations result from the free end conditions for $\{x_2, x_3\}$.

Integrating these equations yields

$$\begin{aligned} \lambda_1 &= bx_1(t_f) \\ \lambda_2 &= bx_1(t_f)(t_f - t) \\ \lambda_3 &= bx_1(t_f)\left[(t_f - t)T - T^2 + T^2 e^{(t-t_f)/T}\right] \end{aligned} \tag{4.19}$$

We conclude that

$$\begin{aligned} u &= -bx_1(t_f)\left[(t_f - t) - T + Te^{(t-t_f)/T}\right] \\ w &= \gamma^{-2} bx_1(t_f)(t_f - t) \end{aligned} \tag{4.20}$$

We can basically repeat the process described by Eqs. (3.45–3.48). First substitute Eqs. (4.20) into Eqs. (4.17), then integrate the resulting differential equations from t_0 to t to get $x_1(t)$, and finally, after evaluating the result for $t = t_f$, solve for $x_1(t_f)$ to get expressions for u and w. The associated MAPLE program is presented in Listing 4.3. The resulting feedback form for u is as follows:

$$u(t) = -(g_1 x_1 + g_2 x_2 + g_3 x_3) \tag{4.21}$$

where

$$\begin{aligned} g_1 &= \frac{1}{\tau^2} \frac{6h^2(h - 1 + e^{-h})}{(6/T^3 b) + [2(1-\gamma^{-2})h^3 + 3 - 6h^2 + 6h - 12he^{-h} - 3e^{-2h}]} \\ g_2 &= \frac{1}{\tau} \frac{6h^2(h - 1 + e^{-h})}{(6/T^3 b) + [2(1-\gamma^{-2})h^3 + 3 - 6h^2 + 6h - 12he^{-h} - 3e^{-2h}]} \\ g_3 &= \frac{6(h - 1 + e^{-h})^2}{(6/T^3 b) + [2(1-\gamma^{-2})h^3 + 3 - 6h^2 + 6h - 12he^{-h} - 3e^{-2h}]} \end{aligned} \tag{4.22}$$

where, as before, $\tau = t_f - t$ and $h = \tau/T$.

Remark: Comparing this result with the one obtained for the simpler target case [Eqs. (3.48)], we notice that the gains here are generally higher because of the $(1 - \gamma^{-2})$ term in the denominator. Recall that in the present case no measurement of the target's acceleration is required.

Consider now the possibility of a conjugate point. Recall that a finite escape time of $g_i(t)$ indicates a conjugate point. We, therefore, need to investigate the existence of positive roots of

$$p(h) = \frac{6}{T^3 b} + [2(1-\gamma^{-2})h^3 + 3 - 6h^2 + 6h - 12he^{-h} - 3e^{-2h}]$$

If $p(h)$ has a positive root h_0, then g_i has a finite escape point at $\tau = Th_0$.

Consider first perfect intercepts (i.e., $b \to \infty$), thus

$$d(h, \gamma) \equiv p_{b\to\infty}(h) = [2(1-\gamma^{-2})h^3 + 3 - 6h^2 + 6h - 12he^{-h} - 3e^{-2h}]$$

Listing 4.3 Gains for nonideal pursuer and ideal target

```
> #listing 4.3
> #l1 is constant and equals c*x1f
> #l2 is linear and equal c*x1f*(tf-t)
> # the following is the remaining costate equation
> #in the sequel T=tau
> sys:=diff(l3(t),t)+c*x1f*(tf-t)-l3(t)/tau;
> dsolve({sys,l3(tf)=0},l3(t));
> assign(");
> simplify(");
> #
> # the following is the state equation for x3(t), namely d(x3)/dt=(x3+u)/T;
> u=-l3/T; x3(0)=a0.
> #
> sys1:=diff(x3(t),t)+x3(t)/tau+l3(t)/(tau^2);
> dsolve({sys1,x3(0)=a0},x3(t));
> assign(");
> simplfy(");
> #
> # state equation for x2, namely: dx2/dt=x3+w; w=l2/beta^2; x2(0)=v0.
> #
> sys2:=diff(x2(t),t)-x3(t)-beta^(-2)*c*x1f*(tf-t);
> dsolve({sys2,x2(0)=v0},x2(t));
> assign(");
> #
> # state equation for x1, namely: dx1/dt=x2; x21(0)=x0.
> #
> sys3:=diff(x1(t),t)-x2(t)
> ;dsolve({sys3,x1(0)=x0},x1(t));
> assign(");
> #
> #evaluating and solving for x1f
> #
> x1(t);
> subs(t=tf,");
> sys5:=x1f-";
> solve(",{x1f});
```

Figure 4.1a is a three-dimensional plot of the function $d(h, \gamma)$ over the domain $\gamma \in [1, 10]$ and $h \in [0, 6]$. It is clear from the Fig. 4.1a that for small γ there exist real roots (hence, conjugate points) for this function in the indicated domain. We will prove that for all $\gamma > 1$ there is a positive h for which $d(h, \gamma) = 0$. First, notice that as $h \to \infty$, we get

$$\frac{d(h, \gamma)}{h^3} \to 2(1 - \gamma^{-2}) > 0$$

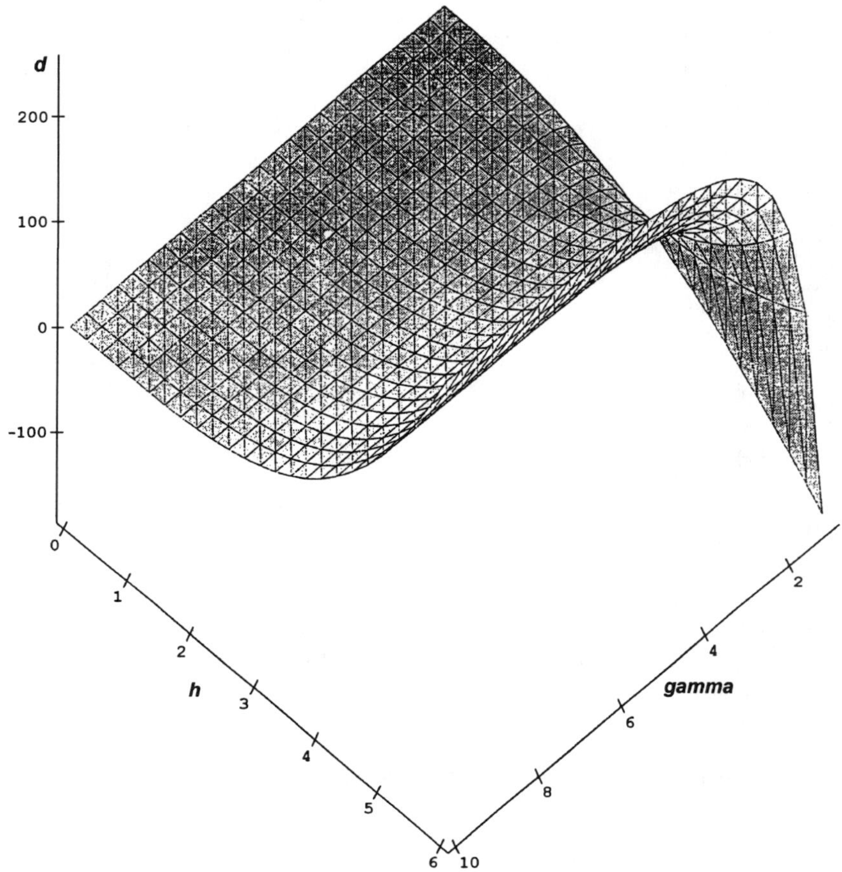

Fig. 4.1a Singularity check.

for all $\gamma > 1$. On the other hand, for $h \to 0$ we get (applying L'Hôpital's rule)

$$\frac{d(h,\gamma)}{h^3} \to -2\gamma^{-2} < 0$$

Hence, a crossover at a conjugate point is inevitable. Notice that the case $\gamma = \infty$ does not suffer from this problem, and indeed Ref. 6 has obtained bounded feedback gains for $t < t_f$. To avoid this undesirable situation, we need to give up perfect intercepts by employing finite b in the performance index. Consequently, the surface of Fig. 4.1a is raised by the amount $6/bT^3$ to avoid a conjugate point. Moreover, the higher we make γ, the less we need to elevate this surface. There is, therefore, a trade-off between b and γ. The physical meaning of this trade-off is quite simple: for more maneuverable targets (low γ) the miss distance increases. Notice, however, that in the ideal missile case a zero miss distance is guaranteed for $\gamma > 1$; this is no longer the case here.

B. Nonideal Target

Assuming first-order lags for both the pursuer and the evader we get the following governing equations in state space:

$$\dot{x} = Ax + Bu + Dw$$

$$A = \begin{bmatrix} 0 & 1 & 0 & 0 \\ 0 & 0 & 1 & 1 \\ 0 & 0 & -\dfrac{1}{T} & 0 \\ 0 & 0 & 0 & -\dfrac{1}{\Theta} \end{bmatrix} \qquad B = \begin{bmatrix} 0 \\ 0 \\ \dfrac{1}{T} \\ 0 \end{bmatrix} \qquad D = \begin{bmatrix} 0 \\ 0 \\ 0 \\ \dfrac{1}{\Theta} \end{bmatrix} \tag{4.23}$$

where Θ is the time constant representing the evader's acceleration control system and w is the acceleration command. The problem is again

$$\min_u \max_w J = \frac{b}{2} x_1^2(t_f) + \frac{1}{2} \int_{t_0}^{t_f} [u^2(t) - \gamma^2 w^2(t)]\, \mathrm{d}t$$

The adjoint equations and the corresponding terminal conditions for this case are as follows:

$$\dot{\lambda}_1 = 0 \qquad \dot{\lambda}_2 = -\lambda_1 \qquad \dot{\lambda}_3 = -\lambda_2 + \frac{\lambda_3}{T} \qquad \dot{\lambda}_4 = -\lambda_2 + \frac{\lambda_4}{\Theta} \tag{4.24}$$

$$u = -\frac{1}{T}\lambda_3 \qquad w = \gamma^{-2} \frac{1}{\Theta} \lambda_4$$

$$\lambda_1(t_f) = b x_1(t_f) \qquad \lambda_2(t_f) = \lambda_3(t_f) = \lambda_4(t_f) = 0$$

Integrating the adjoint equations yields

$$\begin{aligned} \lambda_1 &= b x_1(t_f) \\ \lambda_2 &= b x_1(t_f)(t_f - t) \\ \lambda_3 &= b x_1(t_f)\left[(t_f - t)T - T^2 + T^2 e^{(t-t_f)/T}\right] \\ \lambda_4 &= b x_1(t_f)\left[(t_f - t)\Theta - \Theta^2 + \Theta^2 e^{(t-t_f)/T}\right] \end{aligned} \tag{4.25}$$

We conclude that

$$\begin{aligned} u &= -b x_1(t_f)\left[(t_f - t) - T + T e^{(t-t_f)/T}\right] \\ w &= \gamma^{-2} b x_1(t_f)\left[(t_f - t) - \Theta + \Theta e^{(t-t_f)/T}\right] \end{aligned} \tag{4.26}$$

We can now basically repeat the process described previously. First we substitute Eqs. (4.26) into Eqs. (4.23), then integrate the resulting differential equations to get $x_1(t)$, and finally, after evaluating the result at $t = t_f$, we solve for $x_1(t_f)$ to get expressions for u and w. A derivation of the entire process in MAPLE is given in Listing 4.4.

The resulting feedback form for u is as follows:

$$u(t) = -(g_1 x_1 + g_2 x_2 + g_3 x_3 + g_4 x_4) \tag{4.27}$$

Listing 4.4 Gains for nonideal pursuer and target

```
> #listing 4.4
> #l1 is constant and equals c*x1f
> #l2 is linear and equal c*x1f*(tf−t)
> # the following is the costate equation for l3, namely dl3/dt=l3/T−l2;
> # in the sequel T=tau
> sys:=diff(l3(t),t)+b*x1f*(tf−t)−l3(t)/tau;
> dsolve({sys,l3(tf)=0},l3(t));
> assign(");
> simplify(");
> #
> # the following is the state equation for x3(t), namely d(x3)/dt=(x3+u)/T;
> # u=−l3/T; x3(0)=a0.
> #
> sys1:=diff(x3(t),t)+x3(t)/tau+l3(t)/(tau^2);
> dsolve({sys1,x3(0)=a0},x3(t));
> assign(");
> simplify(");
> #
> # the following is the costate equation for l4, namely dl3/dt=l3/Theta−l2;
> #
> sys:=diff(l4(t),t)+b*x1f*(tf−t)−l4(t)/theta;
> dsolve({sys,l4(tf)=0},l4(t));
> assign(");
> simplify(");
> #
> # the following is the state equation for x4(t), namely #
> d(x4)/dt=(x3+w)/Theta;; w=l4/Theta*beat^2; x3(0)=a0.
> #
> sys1:=diff(x4(t),t)+x4(t)/theta−l4(t)/(beta^2*theta^2);
> dsolve({sys1,x4(0)=aT0},x4(t));
> assign(");
> simplify(");
> #
> # state equation for x2, namely: dx2/dt=x3+x4; x2(0)=v0.
> #
> sys2:=diff(x2(t),t)−x3(t)−x4(t);
> dsolve({sys2,x2(0)=v0},x2(t));
> assign(");
> #
> # state equation for x1, namely: dx1/dt=x2; x1(0)=x0.
> #
> sys3:=diff(x1(t),t)−x2(t);
> dsolve({sys3,x1(0)=x0},x1(t));
> assign(");
> #
> #evaluating and solving for x1f
```

Cont.

Listing 4.4 (Continued)

```
> #
> x1(t);
> subs(t=tf,");
> sys5:=x1f−";
> solve(",{x1f});
> assign(");
> #
> # evaluating the control at t=0
> #
> l3t:=eval(l3(t));
> simplify(");
> u:=−1/T*(");
> u0:=subs(t=0,");
> limit(",beta=infinity);
```

where

$$g_1 = \frac{1}{\tau^2} \times \frac{6h^2(h-1+e^{-h})}{(6/T^3b) + [2(1-\gamma^{-2})h^3 + 3 - 6h^2 + 6h - 12he^{-h} - 3e^{-2h}] + \gamma^{-2}f(\bar{h}, h)}$$

$$g_2 = \frac{1}{\tau} \times \frac{6h^2(h-1+e^{-h})}{(6/T^3b) + [2(1-\gamma^{-2})h^3 + 3 - 6h^2 + 6h - 12he^{-h} - 3e^{-2h}] + \gamma^{-2}f(\bar{h}, h)}$$

$$g_3 = \frac{6(h-1+e^{-h})^2}{(6/T^3b) + [2(1-\gamma^{-2})h^3 + 3 - 6h^2 + 6h - 12he^{-h} - 3e^{-2h}] + \gamma^{-2}f(\bar{h}, h)}$$

$$g_4 = \left(\frac{h}{\bar{h}}\right)^2 \times \frac{6(\bar{h}-1+e^{-\bar{h}})(h-1+e^{-h})}{(6/T^3b) + [2(1-\gamma^{-2})h^3 + 3 - 6h^2 + 6h - 12he^{-h} - 3e^{-2h}] + \gamma^{-2}f(\bar{h}, h)}$$

$$f(\bar{h}, h) = \left[-3\left(\frac{h}{\bar{h}}\right)^3 + 6\left(\frac{h}{\bar{h}}\right)h^2 - 6\left(\frac{h}{\bar{h}}\right)^2 h + 12\left(\frac{h}{\bar{h}}\right)^2 he^{-\bar{h}} + 3\left(\frac{h}{\bar{h}}\right)^3 e^{-2\bar{h}}\right] \tag{4.28}$$

and where

$$h = \tau/T \qquad \bar{h} = \tau/\Theta \tag{4.29}$$

Notice that the pursuer should have estimates of the target's maneuver (as well as its own) to carry out this closed-loop strategy. Notice also that, unlike the one-sided optimal-control approach, the first three gains are different [due to $f(\bar{h}, h)$] than for the ideal target case [e.g., g_3, the feedback gain of the pursuer's own acceleration, is affected by the target's model (in contrast to the one-sided optimal solution)]. As a consequence, the study of conjugate point possibilities is more involved and is better carried out numerically rather than symbolically.

IV. Adding a Constant Maneuver to the Pursuit-Evasion Problem

Recall that in Chapter 3 a constant maneuver was considered, and the pursuing missile used an estimate, based either on measurements or on a predetermined value, of the maneuver to employ proportional feedback [Eq. (3.33)]. As described in Chapter 2, we want to consider cases with a combination of this constant ("dumb") target maneuver and an evasive ("smart") maneuver. Notice that the pursuer is assumed to have estimates for the constant part only.

We begin with the simplest case of an ideal adversary; hence we use the following dynamic equation (with a constant w_0):

$$\dot{x} = Ax + Bu + D(w + w_0)$$
$$A = \begin{bmatrix} 0 & 1 \\ 0 & 0 \end{bmatrix} \qquad B = \begin{bmatrix} 0 \\ 1 \end{bmatrix} \qquad D = \begin{bmatrix} 0 \\ 1 \end{bmatrix} \tag{4.30}$$

For simplicity we assume that the terminal normal relative velocity is of no significance. Therefore the optimal minmax problem is as follows:

$$\min_u \max_w J = \frac{b}{2} x_1^2(t_f) + \frac{1}{2} \int_{t_0}^{t_f} [u^2(t) - \gamma^2 w^2(t)] \, \mathrm{d}t \tag{4.31}$$

Notice that γ now represents the relative maneuverabilty with respect to the smart maneuver only, without considering w_0.

The dynamic equations for the costates remain

$$\dot{\lambda}_1 = 0 \qquad \dot{\lambda}_2 = -\lambda_1 \tag{4.32}$$

and also the minmax operation yields again

$$u = -\lambda_2 \qquad w = \gamma^{-2} \lambda_2 \tag{4.33}$$

From the transversality conditions we infer that

$$\lambda_1 = b x_1(t_f) \qquad \lambda_2 = b x_1(t_f)(t_f - t) \tag{4.34}$$

Substituting $u = -[b x_1(t_f)(t_f - t)]$ and $w = \gamma^{-2}[b x_1(t_f)(t_f - t)]$ into Eqs. (4.30), and then integrating it from t_0 to $t = t_f$, we get the following expression for $x_1(t_f)$:

$$x_1(t_f) = x_1(t_0) + x_2(t_0)(t_f - t_0) + \frac{w_0}{2}(t_f - t_0)^2 - \frac{b}{3}(1 - \gamma^{-2}) x_1(t_f)(t_f - t_0)^3 \tag{4.35}$$

Solving for $x_1(t_f)$ we obtain

$$x_1(t_f) = \frac{1/b}{(1/b) + \frac{1}{3}(1-\gamma^{-2})(t_f - t_0)^3}\left[x_1(t_0) + x_2(t_0)(t_f - t_0) + \frac{w_0}{2}(t_f - t_0)^2\right] \tag{4.36}$$

Since $u(t_0) = -[bx_1(t_f)(t_f - t_0)]$, by omitting the subindex in t_0, we obtain the following linear feedback form:

$$u(t) = \frac{-1}{(1/b) + \frac{1}{3}(1-\gamma^{-2})(t_f - t)^3}\left[x_1(t)(t_f - t) + x_2(t)(t_f - t)^2 + \frac{w_0}{2}(t_f - t)^3\right] \tag{4.37}$$

Listing 4.5 presents a MAPLE routine to carry out the derivation of this problem. For perfect interception $b \to \infty$, and we get

$$u = \frac{-3V_c\dot{\lambda}}{(1-\gamma^{-2})} + \frac{-3}{2}\frac{w_0}{(1-\gamma^{-2})} \tag{4.38}$$

[see Eq. (3.14)]. Notice first that for $\gamma \to \infty$, the one-sided optimal solution is obtained. Also notice that the solution is not a simple superposition of the smart and dumb target cases (i.e., the optimal and the constant evasive maneuvering targets). Rather, the feedback from w_0 is affected by the fact that the target performs an additional optimal maneuver.

Consider now the nonideal pursuer with the same assumptions concerning the evasive maneuvers. Thus we write

$$\dot{x} = Ax + Bu + D(w + w_0)$$

$$A = \begin{bmatrix} 0 & 1 & 0 \\ 0 & 0 & 1 \\ 0 & 0 & -\frac{1}{T} \end{bmatrix} \qquad B = \begin{bmatrix} 0 \\ 0 \\ \frac{1}{T} \end{bmatrix} \qquad D = \begin{bmatrix} 0 \\ 1 \\ 0 \end{bmatrix} \tag{4.39}$$

The pursuit-evasion problem is as before:

$$\min_u \max_w J = \frac{b}{2}x_1^2(t_f) + \frac{1}{2}\int_{t_0}^{t_f} [u^2(t) - \gamma^2 w^2(t)]\,\mathrm{d}t \tag{4.40}$$

The adjoint equations are again of the form:

$$\begin{gathered} \dot{\lambda}_1 = 0 \qquad \dot{\lambda}_2 = -\lambda_1 \qquad \dot{\lambda}_3 = -\lambda_2 + \frac{\lambda_3}{T} \\ u = -\frac{1}{T}\lambda_3 \qquad w = \gamma^{-2}\lambda_2 \end{gathered} \tag{4.41}$$

with the terminal conditions

$$\lambda_1(t_f) = bx_1(t_f) \qquad \lambda_2(t_f) = 0 \qquad \lambda_3(t_f) = 0 \tag{4.42}$$

Listing 4.5 Gains for ideal pursuer and target with additional constant maneuver

```
> #listing 4.5
> # l1(t) is constant and equals b*x1f.
> # the following is the remaining costate equation for l2(t), namely d(l2)/dt=−l1
> sys:=diff(l2(t),t)+b*x1f;
> dsolve({sys,l2(tf)=0},l2(t));
> assign(");
> simplify(");
> #
> # the following is the state equation for x2(t), namely d(x2)/dt=u+w;u=−l2
> # w=l2/beta^2+w0
> #
> sys1:=diff(x2(t),t)+(1−beta^(−2))*l2(t)−w0;
> dsolve({sys1,x2(0)=v0},x2(t));
> assign(");
> simplify(");
> #
> # the following is the state equation of x1(t), namely d(x1)/dt=x2.
> #
> sys3:=diff(x1(t),t)−x2(t);
> dsolve({sys3,x1(0)=x0},x1(t));
> assign(");
> #
> # evaluating and solving for x1f
> #
> x1(t);
> y1:=subs(t=tf,");
> sys5:={x1f−y1};
> solve(", {x1f});
> assign(");
> #
> # evaluating the control u=−l2 at t=0
> #
> u:=−eval(l2(t));
> u0:=subs(t=0,");
> simplify(");
> #limit value for perfect intercept
> limit(",b=infinity);
```

These equations have been integrated before to yield

$$
\begin{aligned}
\lambda_1 &= bx_1(t_f) \\
\lambda_2 &= bx_1(t_f)(t_f - t) \\
\lambda_3 &= bx_1(t_f)\left[(t_f - t)T - T^2 + T^2 e^{(t-t_f)/T}\right]
\end{aligned}
\tag{4.43}
$$

We conclude that

$$u = -bx_1(t_f)\left[(t_f - t) - T + Te^{(t-t_f)/T}\right]$$
$$w = \gamma^{-2}bx_1(t_f)(t_f - t) \tag{4.44}$$

We now repeat the same process for finding u in feedback form. First we substitute Eqs. (4.44) into Eqs. (4.39), then integrate the resulting differential equations to get $x_1(t)$, and finally, after evaluating the result at $t = t_f$, we solve for $x_1(t_f)$ to get expressions for u and w. Listing 4.6 contains a MAPLE listing of the procedure. We obtain the following expression for u:

$$u(t) = -(g_1x_1 + g_2x_2 + g_3x_3 + g_4w_0)$$

where

$$g_1 = \frac{1}{\tau^2}\frac{6h^2(h - 1 + e^{-h})}{(6/T^3b) + [2(1-\gamma^{-2})h^3 + 3 - 6h^2 + 6h - 12he^{-h} - 3e^{-2h}]}$$

$$g_2 = \frac{1}{\tau}\frac{6h^2(h - 1 + e^{-h})}{(6/T^3b) + [2(1-\gamma^{-2})h^3 + 3 - 6h^2 + 6h - 12he^{-h} - 3e^{-2h}]}$$

$$g_3 = \frac{6(h - 1 + e^{-h})^2}{(6/T^3b) + [2(1-\gamma^{-2})h^3 + 3 - 6h^2 + 6h - 12he^{-h} - 3e^{-2h}]} \tag{4.45}$$

$$g_4 = \frac{3h^2(h - 1 + e^{-h})}{(6/T^3b) + [2(1-\gamma^{-2})h^3 + 3 - 6h^2 + 6h - 12he^{-h} - 3e^{-2h}]}$$

where $\tau = t_f - t$ and $h = \tau/T$.

Notice that these gains differ from the one-sided optimal control gains [Eqs. (3.48)] only by the term $(1 - \gamma^{-2})$, which converges to 1 as $\gamma \to \infty$.

V. General Solution for Ideal Adversaries

At this point we can conclude by presenting the general solution to the interception problem, in the sense that both smart (unmeasured) and dumb (measured) maneuvers are taken into account, as are velocity considerations and the miss distances. For simplicity, we restrict the discussion to ideal adversaries. The general formulation for a nonideal pursuer will be discussed in Chapter 5 in the context of sensitivity reduction.

Therefore, the problem formulation is

$$\dot{x} = Ax + Bu + D(w + w_0)$$

$$A = \begin{bmatrix} 0 & 1 \\ 0 & 0 \end{bmatrix} \qquad B = \begin{bmatrix} 0 \\ 1 \end{bmatrix} \qquad D = \begin{bmatrix} 0 \\ 1 \end{bmatrix} \tag{4.46}$$

Listing 4.6 Gains for nonideal pursuer and ideal target with additional constant maneuver

```
> #listing 4.6
> # l1 is constant and equals c*x1f
> # l2 is linear and equal c*x1f*(tf−t)
> # the following is the remaining costate equation
> #
> sys:=diff(l3(t),t)+b*x1f*(tf−t)−l3(t)/T;
> dsolve({sys,l3(tf)=0},l3(t));
> assign(");
> simplify(");
> #
> # the following is the state equation for x3(t), namely d(x3)/dt=(x3+u)/T;
> u=−l3/T; x3(0)=a0.
> #
> sys1:=diff(x3(t),t)+x3(t)/T+l3(t)/T^2;
> dsolve({sys1,x3(0)=a0},x3(t));
> assign(");
> simplify(")
> #
> # state equation for x2, namely: dx2/dt=x3+w; w=l2/beta^2+w0; x2(0)=v0.
> #
> sys2:=diff(x2(t),t)−x3(t)−w0−beta^(−2)*b*x1f*(tf−t);
> dsolve({sys2,x2(0)=v0},x2(t));
> assign(");
> #
> # state equation for x1, namely: dx1/dt=x2; x21(0)=x0.
> #
> sys3:=diff(x1(t),t)−x2(t);
> dsolve({sys3,x1(0)=x0},x1(t));
> assign(");
> #
> #evaluating and solving for x1f and evaluating the control u at t=0
> #
> x1(t);
> subs(t=tf,");
> sys5:=x1f−";
> solve(",x1f);
> assign(");
> l3t:=eval(l3(t));
> simplify(");
> u:=−1/T*(");
> u0:=subs(t=0,");
```

and the performance index is

$$\min_u \max_w J = \frac{b}{2}x_1^2(t_f) + \frac{c}{2}x_2^2(t_f) + \frac{1}{2}\int_{t_0}^{t_f} [u^2(t) - \gamma^2 w^2(t)]\, \mathrm{d}t \tag{4.47}$$

The solution is

$$\begin{aligned} u &= -(g_1 x_1 + g_2 x_2 + g_3 w_0) \\ g_1 &= \frac{b\gamma^2\tau\left[\frac{1}{2}\tau c(\gamma^2-1)+\gamma^2\right]}{\tau^4(cb/12)(\gamma^2-1)^2 + [\tau c + \tau^3(b/3)]\gamma^2(\gamma^2-1)+\gamma^4} \\ g_2 &= \frac{\frac{1}{3}\gamma^2[\tau^3 cb(\gamma^2-1)+3(\tau^2 b + c)\gamma^2]}{\tau^4(cb/12)(\gamma^2-1)^2 + [\tau c + \tau^3(b/3)]\gamma^2(\gamma^2-1)+\gamma^4} \\ g_3 &= \frac{\frac{1}{12}\gamma^2[\tau^4 cb(\gamma^2-1)] + [\tau^3 b/2 - c\tau]\gamma^4}{\tau^4(cb/12)(\gamma^2-1)^2 + [\tau c + \tau^3(b/3)]\gamma^2(\gamma^2-1)+\gamma^4} \\ \tau &= t_f - t \end{aligned} \tag{4.48}$$

(using the same technique as before, see Listing 4.7).

The most interesting limit cases are 1) perfect intercept (i.e., when $c = 0$ and $b \to \infty$) then

$$g_1 = \frac{3}{\tau^2(1-\gamma^{-2})} \qquad g_2 = \frac{3}{\tau(1-\gamma^{-2})} \qquad g_3 = \frac{3}{2(1-\gamma^{-2})} \tag{4.49}$$

2) perfect rendezvous (i.e., when $b \to \infty$ and $c \to \infty$, then

$$g_1 = \frac{6}{\tau^2(1-\gamma^{-2})} \qquad g_2 = \frac{4}{\tau(1-\gamma^{-2})} \qquad g_3 = \frac{1}{(1-\gamma^{-2})} \tag{4.50}$$

VI. Numerical Examples

The following numerical results were obtained using the previously described MATLAB-based simulation program. For the differential game results finite values were used for the design parameter γ.

A. Nonideal Pursuer and Ideal Evader—The Effect of γ and Finite Miss-Distance Weighting

In this section, the theory for the case of a nonideal pursuer (modeled by a single time constant) and an ideal evader (whose acceleration is not measured by the pursuer), will be demonstrated. Two guidance laws will be compared with respect to their performance in the presence of target maneuvers.

1) A modified version (MEL with no target acceleration measurements, or MELN) of the MELs since the MEL also uses target acceleration measurements, one has to decide how to modify the MEL when such measurements are not available. A possible choice is to assume zero target acceleration. The latter choice leads to the MELN. Note that another possible interpretation for MELN is stochastic. Accordingly, the target's normal acceleration—under the design assumptions—is white noise. The resulting guidance law, as in the previously given SGL, is again MELN.

Listing 4.7 Gains for the general case of ideal adversaries

```
> #listing 4.7
> # l1(t) is constant and equals b*x1f.
> # the following is the remaining costate equation for l2(t), namely d(l2)/dt=−l1.
> #
> sys:=diff(l2(t),t)+b*x1f;
> dsolve({sys,l2(tf)=c*x2f},l2(t));
> assign(");
> simplify(");
> #
> # the following is the state equation for x2(t), namely d(x2)/dt=u+w;u=−l2
> # w=l2/beta^2+w0
> #
> sys1:=diff(x2(t),t)+(1−beta^(−2))*l2(t)−w0;
> dsolve({sys1,x2(0)=v0},x2(t));
> assign(");
> simplify(");
> #
> # the following is the state equation of x1(t), namely d(x1)/dt=x2.
> #
> sys3:=diff(x1(t),t)−x2(t);
> dsolve({sys3,x1(0)=x0},x1(t));
> assign(");
> #
> # evaluating and solving for x1f and x2f
> #
> x1(t);
> y1:=subs(t=tf,");
> x2(t);
> y2:=subs(t=tf,");
> sys5:={x1f−y1,x2f−y2};
> solve(", {x1f,x2f});
> assign(");
> # evaluating the control u=−l2 at t=0
> #
> u:=−eval(l2(t));
> u0:=subs(t=0,");
> simplify(");
> #limit value for perfect intecept
> limit(",b=infinity);
> limit(",c=infinity);
```

2) Another possible choice, when target acceleration measurements are not available, is to assume that the target accelerations are unknown but of finite energy. Such an assumption leads to the differential game approach that was applied to the guidance problem in Sec. III.A. As we will see, this assumption makes the guidance law [Eqs. (4.22)] somewhat robust against the unmeasured maneuvers of the target. This guidance law uses the same measurements as MELN and in fact, the latter is obtained from the former in the limit where $\gamma \to \infty$. We, therefore, refer to both guidance laws as MELN with an indication about the parameter γ to distinguish between the two.

We continue with the assumptions that the pursuer's time constant is 0.5 s, and that the interception conflict lasts for 3 s and is assumed to be a tail chase or head-on pursuit. The closing velocity is 300 m/s.

We consider two types of target maneuvers, namely, a 3 g step in target (evader) acceleration and a sinusoidal evasive maneuver of 3 g of magnitude and of varying frequency. The simulation results are classified in this section according to the maneuver type. The results are illustrated by a comparison of the graphs of relative separation, relative angle (i.e., the quotient of the relative velocity and closing velocity in degrees), commanded pursuer acceleration, and the control effort: $J_u(t) = \int_0^t u^2(\tau)\,d\tau$.

The synthesis of guidance gains using a differential game approach, unlike the case of a one-sided LQ optimization problem, imposes a new difficulty: One has to decide on the miss-distance weighting b and disturbance (maneuvers) attenuation factor γ, so that no conjugate point will appear within the conflict duration. To decide on these parameters, a conjugate point analysis for the specific conflict parameters has to be performed. The results of such an analysis are depicted in Fig. 4.1b. It is seen there that for small $1/b$ (i.e., large miss-distance weighting) the minimum achievable γ is large. Note that, in fact, the minimum achievable γ becomes unbounded as $1/b$ tends to zero, since (see Sec. III.A) a zero miss distance for unknown finite energy maneuvers is unrealistic. As $1/b$ becomes smaller (i.e., larger miss distances are allowed), the minimum achievable γ also becomes smaller. The parameters b and γ are, therefore, design parameters. One has to choose a point $\{1/b, \gamma\}$ on Fig. 4.1b and check whether the performance is satisfactory. The answer may be different for different error sources.

1. Step of 3 g in Target Acceleration

A first step in the comparison is to choose $1/b$ and γ. Note that if γ is chosen to be infinite, then there is no lower bound on $1/b$. Therefore, for infinite γ arbitrarily small miss distances can be achieved; however, small miss distances will involve the use of large pursuer maneuvers. On the other hand, in the case of finite γ the choice of $1/b$ is limited, in our case, by the results in Fig. 4.1b. A fair comparison between an infinite and finite γ will be to use pursuer maneuvers of similar magnitudes against a specific target maneuver. We choose our test maneuver to be a step of 3 g in target acceleration. We also consider the case of $1/b = 10^{-4}$ for infinite γ, and for $\gamma = 4.75$, which is slightly larger than the minimum possible γ indicated by Fig. 4.1b for this $1/b$. We note that at this stage no hard limit is applied to the pursuer's acceleration. In Fig. 4.2 the relative separation for MELN first as $\gamma \to \infty$ and then as $\gamma = 4.75$ are depicted. Both guidance laws show similar results, but the miss distance with the MELN

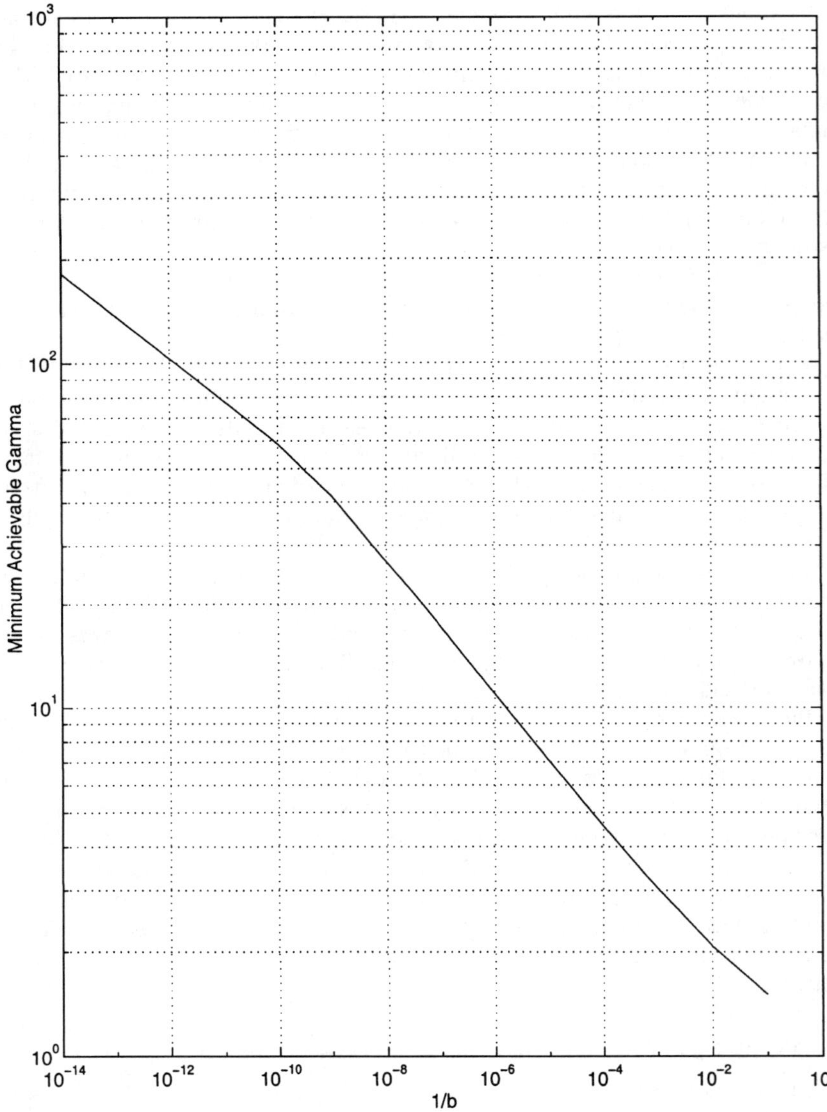

Fig. 4.1b Singularity check.

with $\gamma = 4.75$ is smaller (about 0.5 m compared to about 1 m). The results in Fig. 4.3 indicate that with MELN with $\gamma = 4.75$, a pick maneuver of 17 g is required as opposed to 22 g with MELN as $\gamma \to \infty$. This fact is also exhibited in Fig. 4.4 where the control effort is depicted. These results indicate that the choice of $1/b = 10^{-4}$ allows fair comparison between the $\gamma \to \infty$ and $\gamma = 4.75$ cases.

Since we did not impose any hard limits on the pursuer's maneuverability, the pursuit-evasion conflict is linear and an adjoint analysis (see Chapter 2) can be

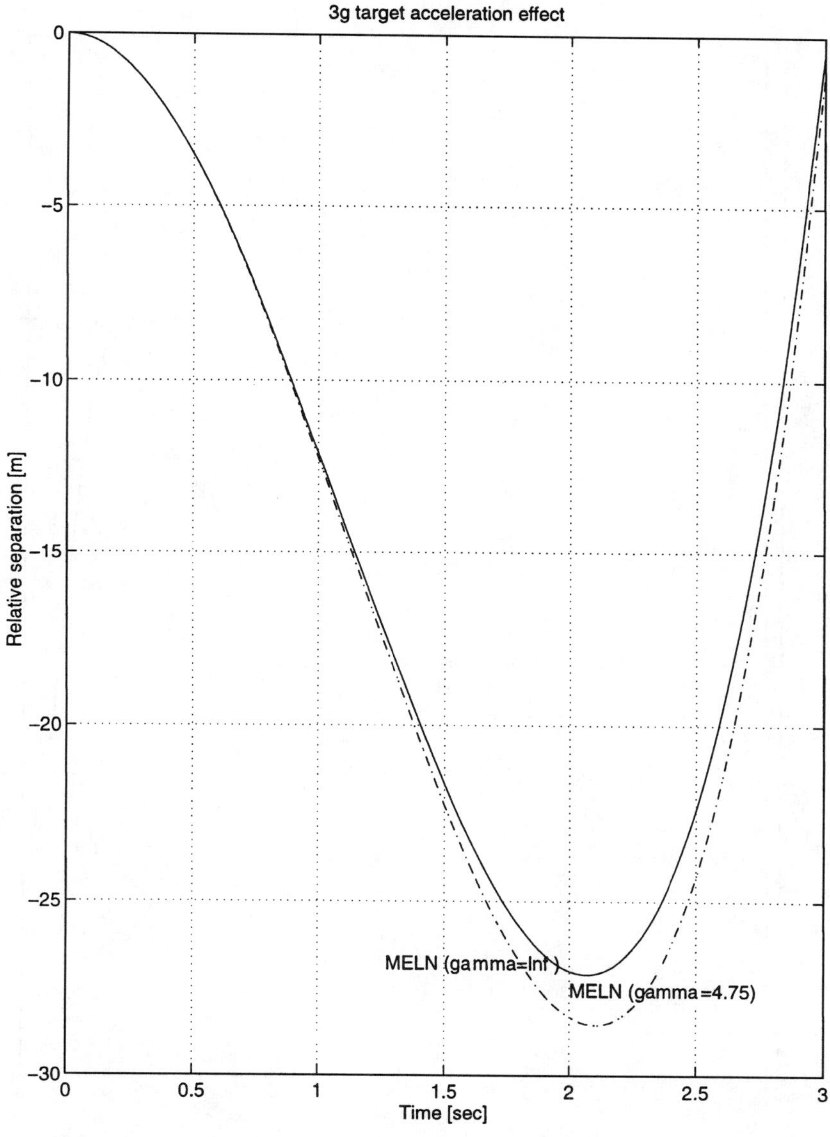

Fig. 4.2 Relative separation due to constant target maneuver.

applied. The results of an adjoint simulation for a step of 3 g in target acceleration are depicted in Fig. 4.5. In contrast to Figs. 4.2–4.4 where only the results for a single conflict duration of 3 s were shown, in Fig. 4.5 we show the whole range of time-to-go (0–3 s) at the start of target maneuvers. The results in Fig. 4.5 are obviously in favor of the lower γ (i.e., $\gamma = 4.75$). For a time-to-go larger than about 0.3 s, the low γ leads to miss distances that are about twice as small as those for the infinite γ.

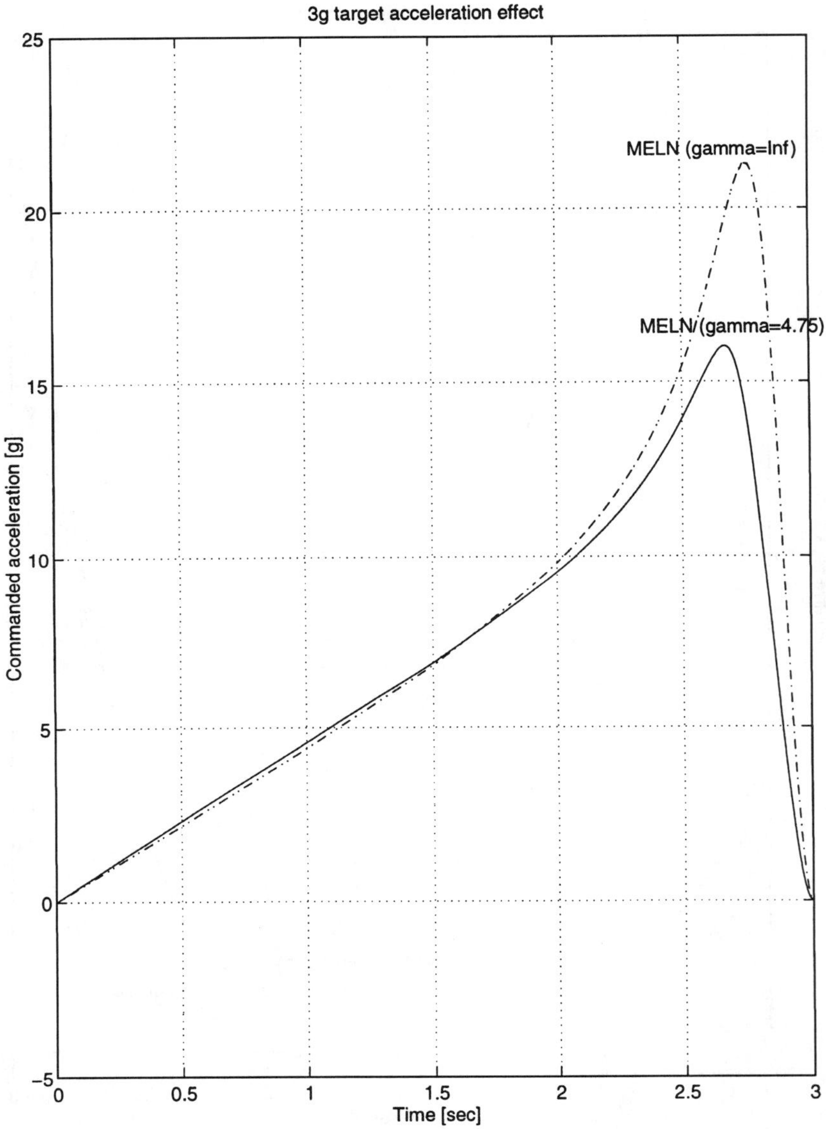

Fig. 4.3 Missile acceleration command due to constant target maneuver.

2. *Heading Error of 10 Deg*

Although the MEL was designed for target maneuvers, we saw in Chapter 3 that it is also effective in the presence of nonzero initial conditions. Therefore, we study the effect of γ on MELN/MEL for heading error with the same problem parameters as in Sec. VI.A.1, where we focused on the case $1/b = 10^{-4}$. In Figs. 4.6–4.8 the relative separation, commanded pursuer acceleration, and control effort are

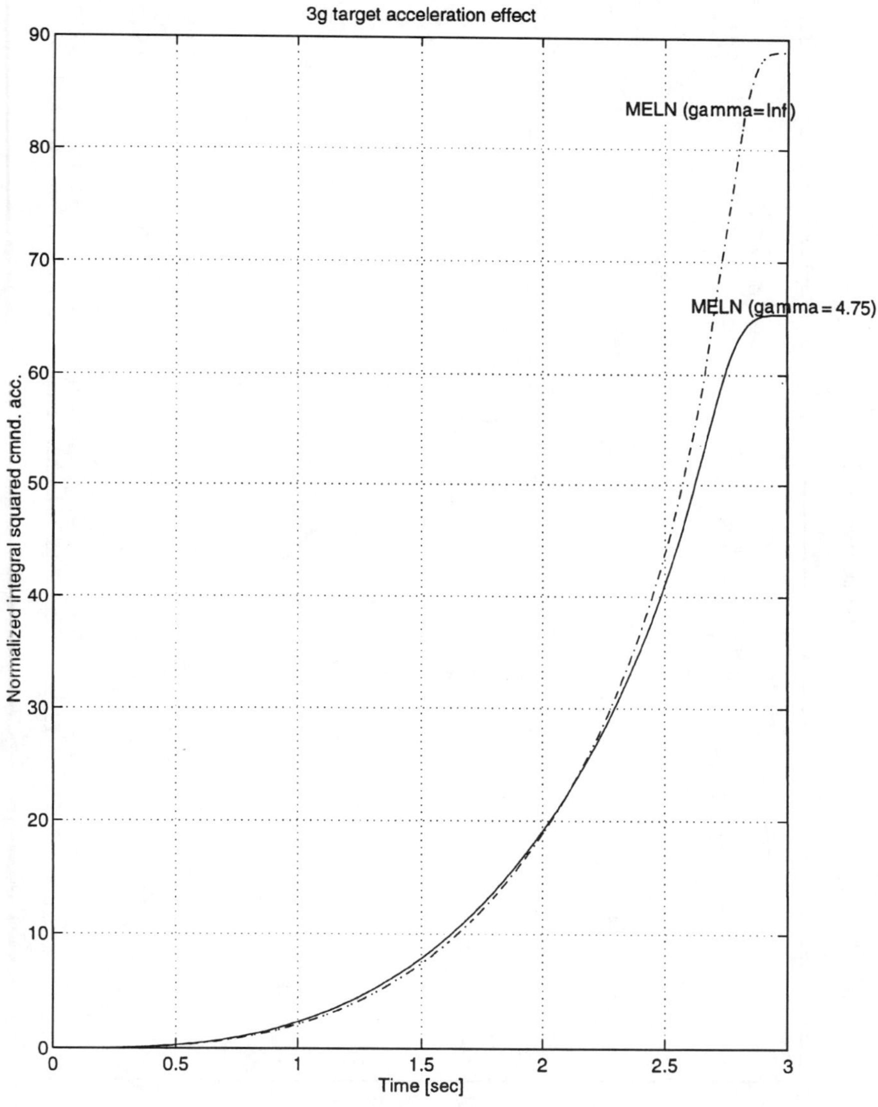

Fig. 4.4 Control effort due to constant target maneuver.

depicted for a conflict duration of 3 s. For this case the MELN with $\gamma = 4.75$ is inferior, requiring slightly larger pursuer maneuvers. An advantage of using MELN with $\gamma = 4.75$ over MELN with $\gamma \to \infty$ is obtained only for shorter duration. This fact is illustrated in Fig. 4.9 where the miss distance due to a heading error with various conflict duration was obtained from an adjoint simulation. (Recall that in the heading error plot we mean t_f when we write time-to-go; this is a slight intentional abuse of notations, see the remark in Sec. V.C.2 of Chapter 3). To complete the analysis we present in Fig. 4.10 the commanded pursuer acceleration in the

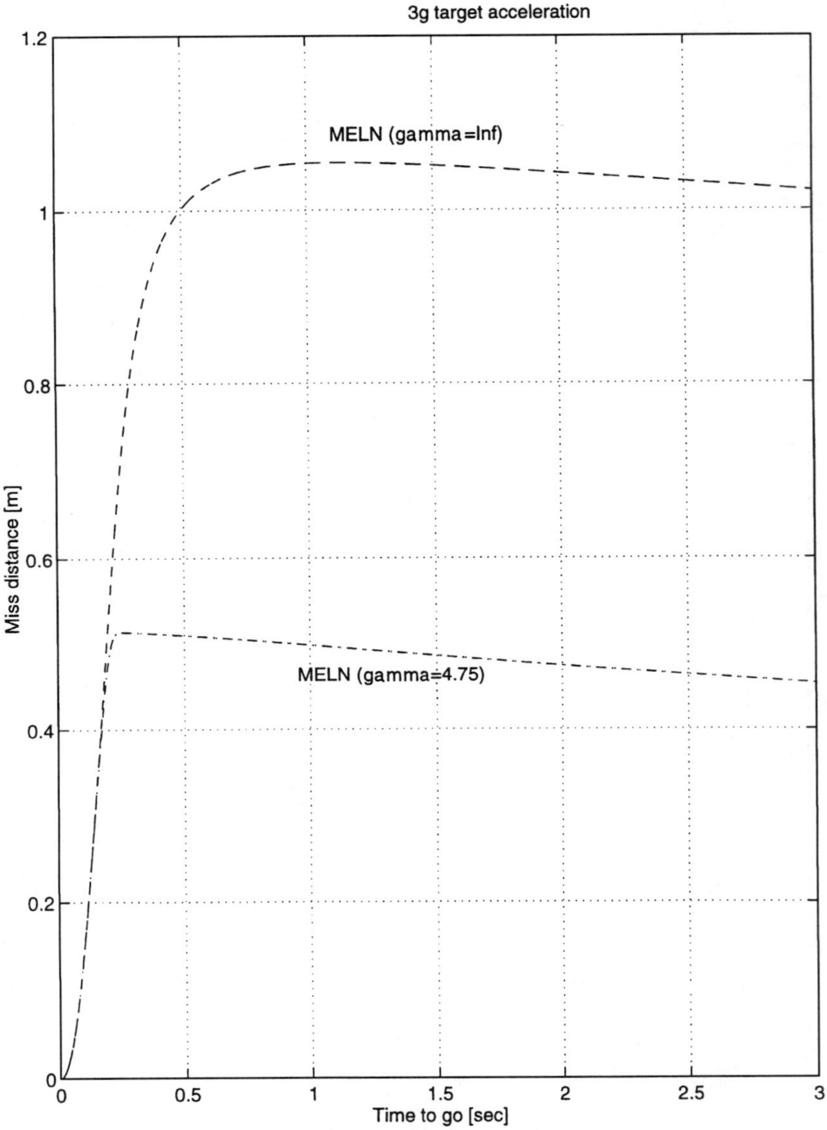

Fig. 4.5 Miss distance due to constant target maneuver.

case of a conflict duration of 0.5 s. We notice that the lower miss distance with MELN having $\gamma = 4.75$ is achieved at the cost of larger initial acceleration commands. MELN with $\gamma = 4.75$ requires 250 g compared to 120 g for MELN with $\gamma \to \infty$ and they yield miss-distance values of about 0.04 m and about 0.64 m, respectively. However, because of the 0.5 s time constant, the actual acceleration (see Fig. 4.11) with the MELN with $\gamma = 4.75$ is somewhat smaller than the MELN with $\gamma \to \infty$.

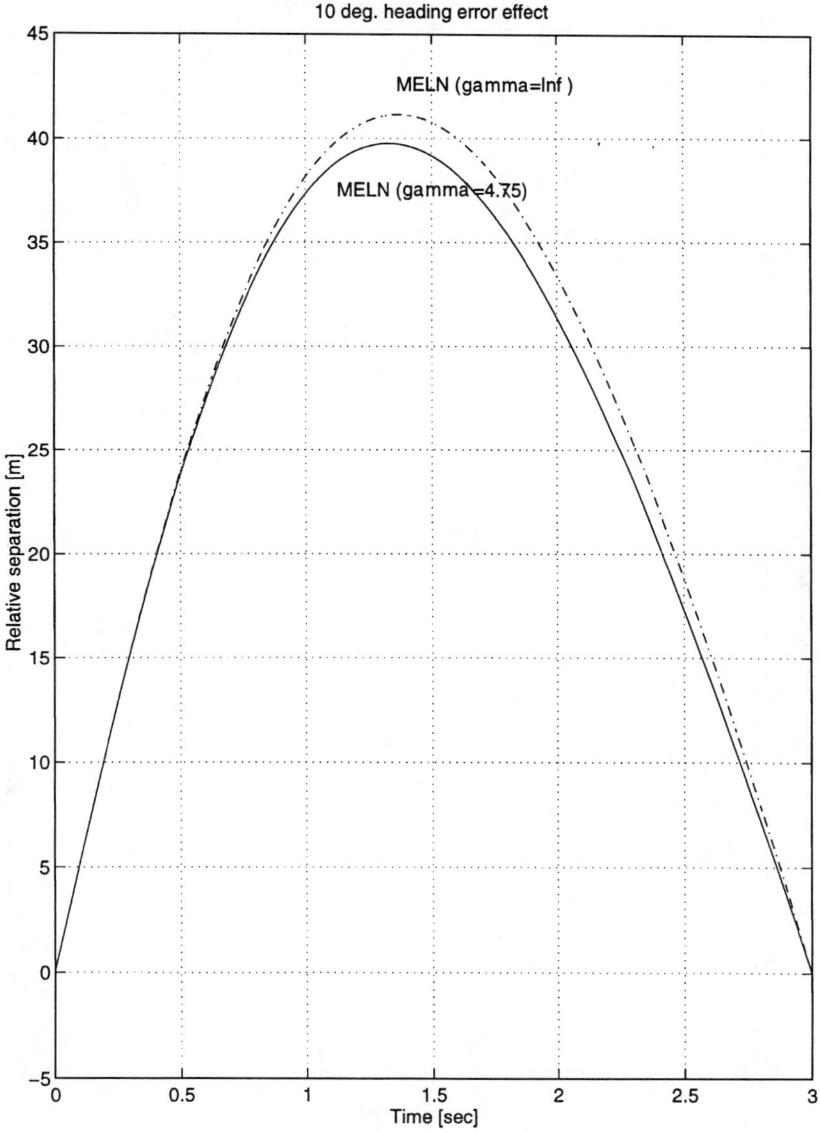

Fig. 4.6 Relative separation due to heading error.

3. Sinusoidal Target Acceleration

Although we use a linearized model of the pursuit-evasion conflict when we consider sinusoidal target maneuvers, we impose (as in Chapter 3) a commanded acceleration limiter of 15 g, so that a more realistic comparison of the guidance laws is allowed. The sinusoidal target accelerations considered are in the frequencies $\{0.1, 0.5, 2, 5, 10\}$ rad/s. A couple of miss-distance weightings are considered, namely $1/b = 10^{-4}$ and $1/b = 0.01$. For each case a nearly minimal γ (Fig. 4.1b)

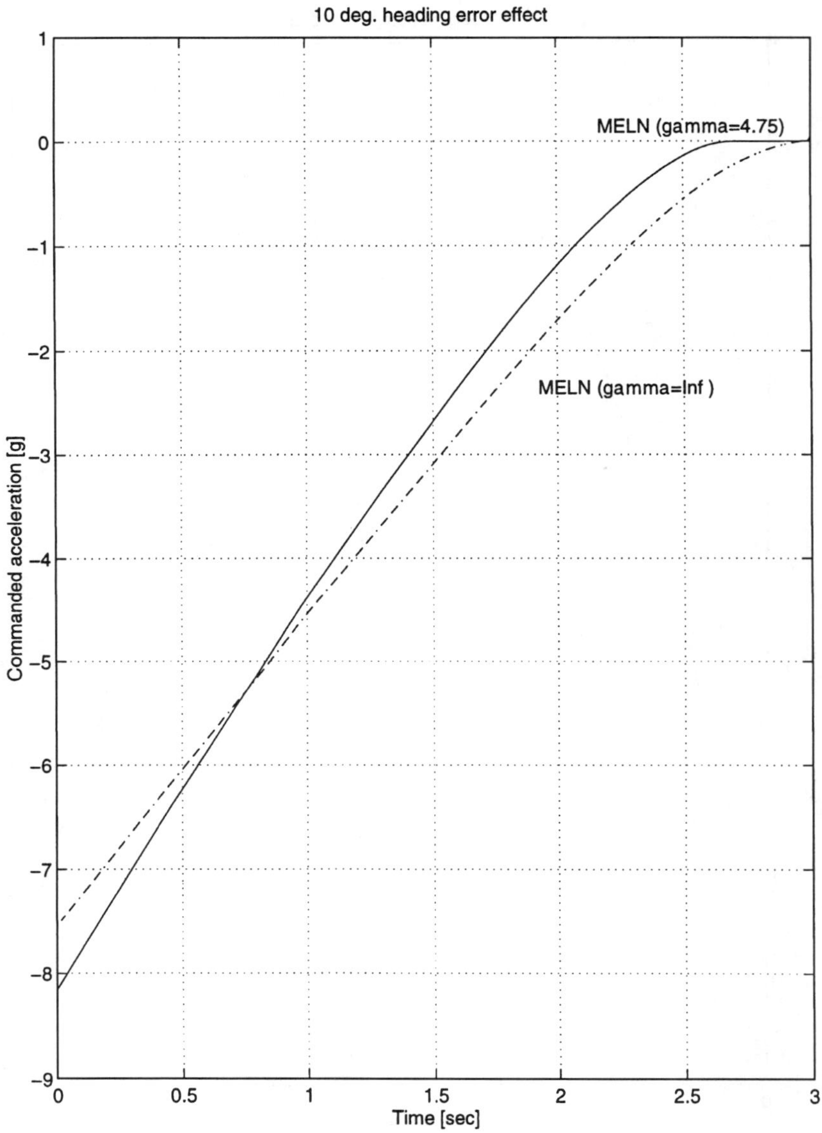

Fig. 4.7 Missile acceleration command due to heading error.

will be taken; for $1/b = 10^{-4}$ it is $\gamma = 4.75$, while for $1/b = 0.01$ it is $\gamma = 2.1$. Since the miss-distance values with sinusoidal target maneuvers depend on the initial phase, we consider for each frequency eight equally spaced initial phase angles between -180 and 180 deg, and we take the largest miss distance in all tested phases as the miss distance for this frequency. In Fig. 4.12 the miss distance for $1/b = 10^{-4}$ and $\gamma = 4.75$ are compared to those with the same $1/b = 10^{-4}$

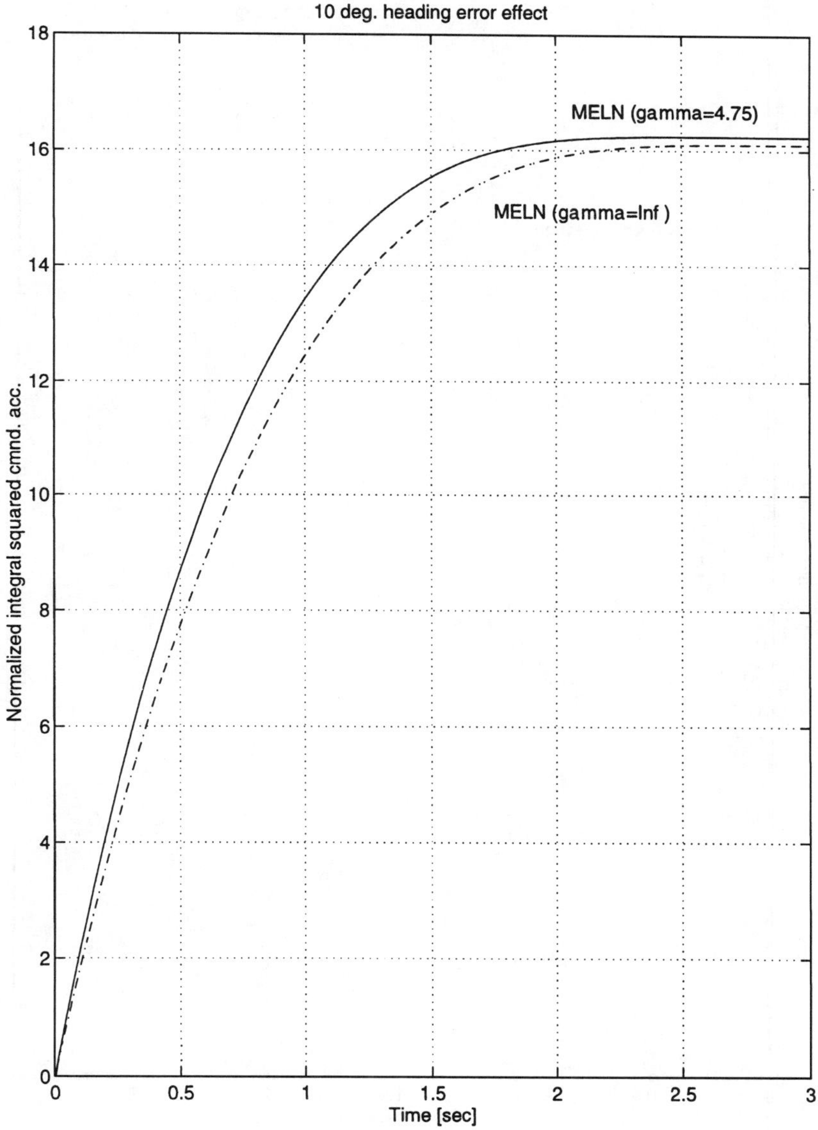

Fig. 4.8 Control effort due to heading error.

but with $\gamma \to \infty$. The miss-distance values are plotted versus frequency and the resulting graphs resemble Bode diagrams. The MELN where $\gamma \to \infty$ has a low-pass, filter-like response with miss distances varying from about 1.1 m to 0.4 m, while the MELN where $\gamma = 4.75$ results in considerably smaller miss-distance values that are almost constant over all frequencies (about 0.4 m). The results in Fig. 4.13 of the $1/b = 0.01$ for $\gamma = 2.1$ and $\gamma \to \infty$ of the MELN show about the same phenomena but with larger miss distances. When $\gamma \to \infty$ the miss distances

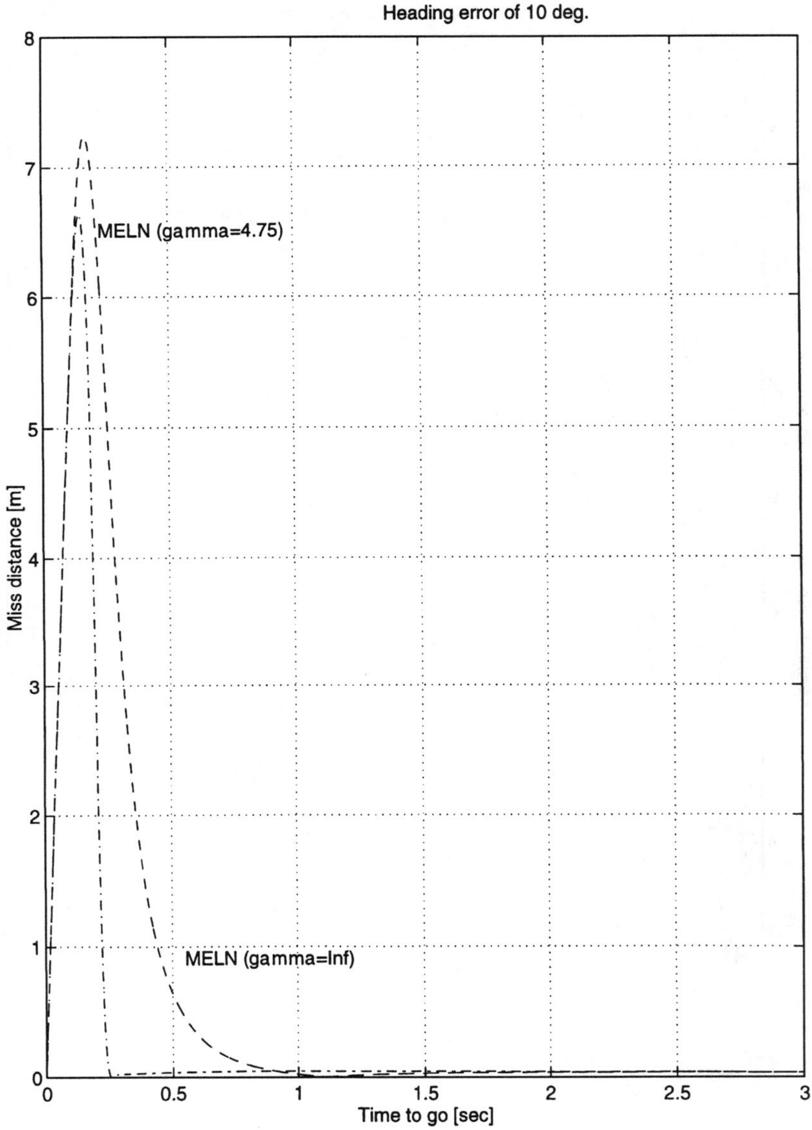

Fig. 4.9 Miss distance due to heading error.

vary between about 8 m and 0.4 m, while with the $\gamma = 2.1$ they are about 4 m until 5 rad/s and then they drop off to about 0.4 m for larger frequencies. The larger miss-distance values with $1/b = 0.01$ are to be expected.

Remark: This phenomenon of almost constant response over all frequencies in the minimum gamma case resembles a phenomenon that is known from the literature on frequency-domain, worst-case design[7] and is often referred to as the equalizing property.

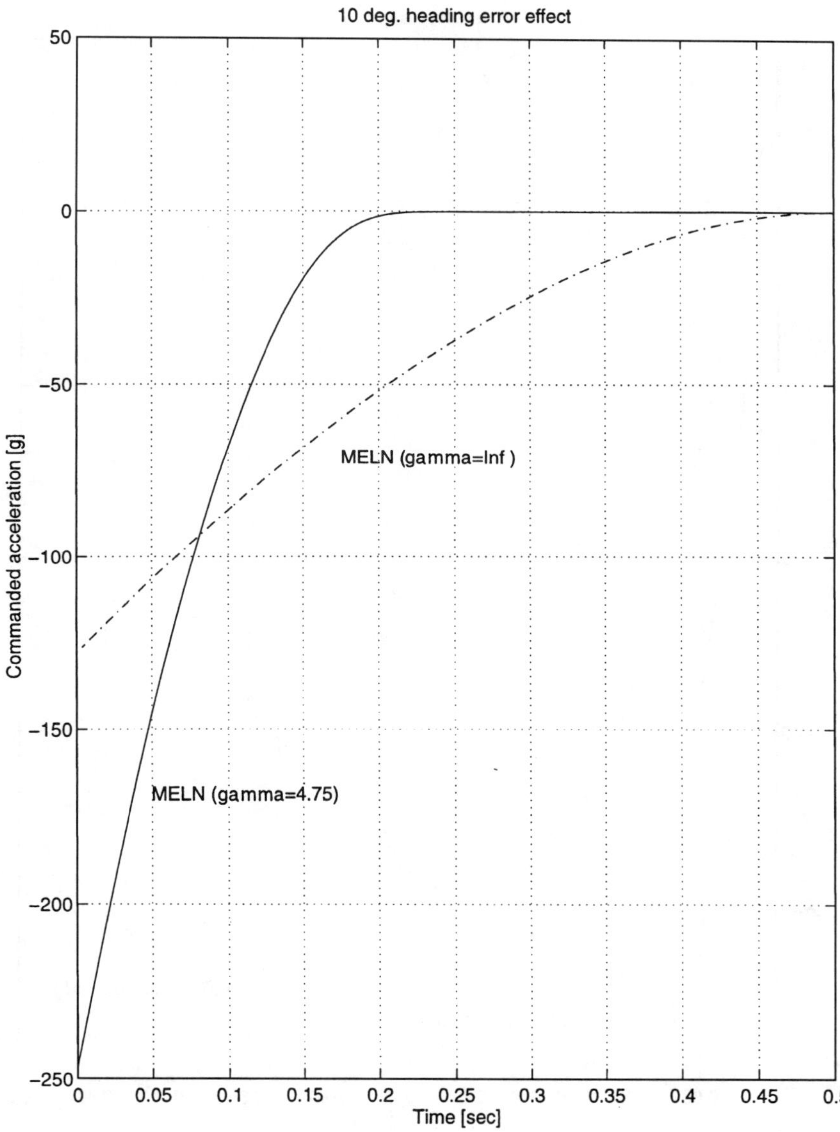

Fig. 4.10 Missile acceleration command due to heading error.

B. Nonideal Pursuer and Evader

In this section, the theory for the case of nonideal pursuer and evader that are both modeled by single time constants will be numerically demonstrated. In contrast to the previous section, we will assume that the evader's acceleration is available to the pursuer. Four guidance laws will be compared.

APN: Recall that this is the extension of the PN guidance law that also uses feedback from the evader's acceleration [Eq. (3.34)]. We consider the APN with a

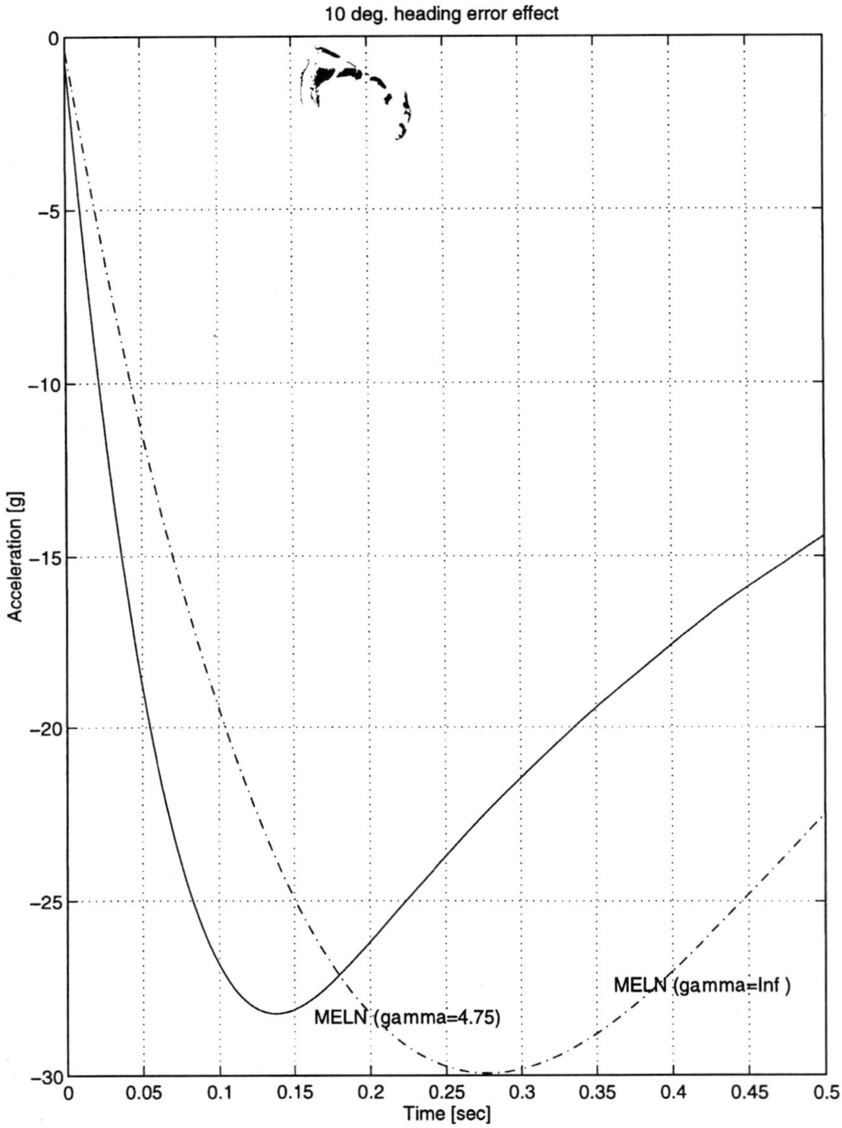

Fig. 4.11 Missile acceleration due to heading error.

navigation constant of 3, which emerges when both time constants of pursuer and evader are ignored in the guidance law synthesis process.

MEL ($\gamma \to \infty$): This is the previously discussed generalized version (with finite miss-distance weighting) of the Zarchan guidance law [Eqs. (3.48)]. Its synthesis procedure ignores the evader's time constant. Moreover, the $\gamma \to \infty$ choice means that the evader is assumed to be forced to constant maneuvers. We expect its Achilles' tendon to be small evader time constants with quickly varying acceleration commands for the evader. For the numerics, we choose $1/b = 0.001$.

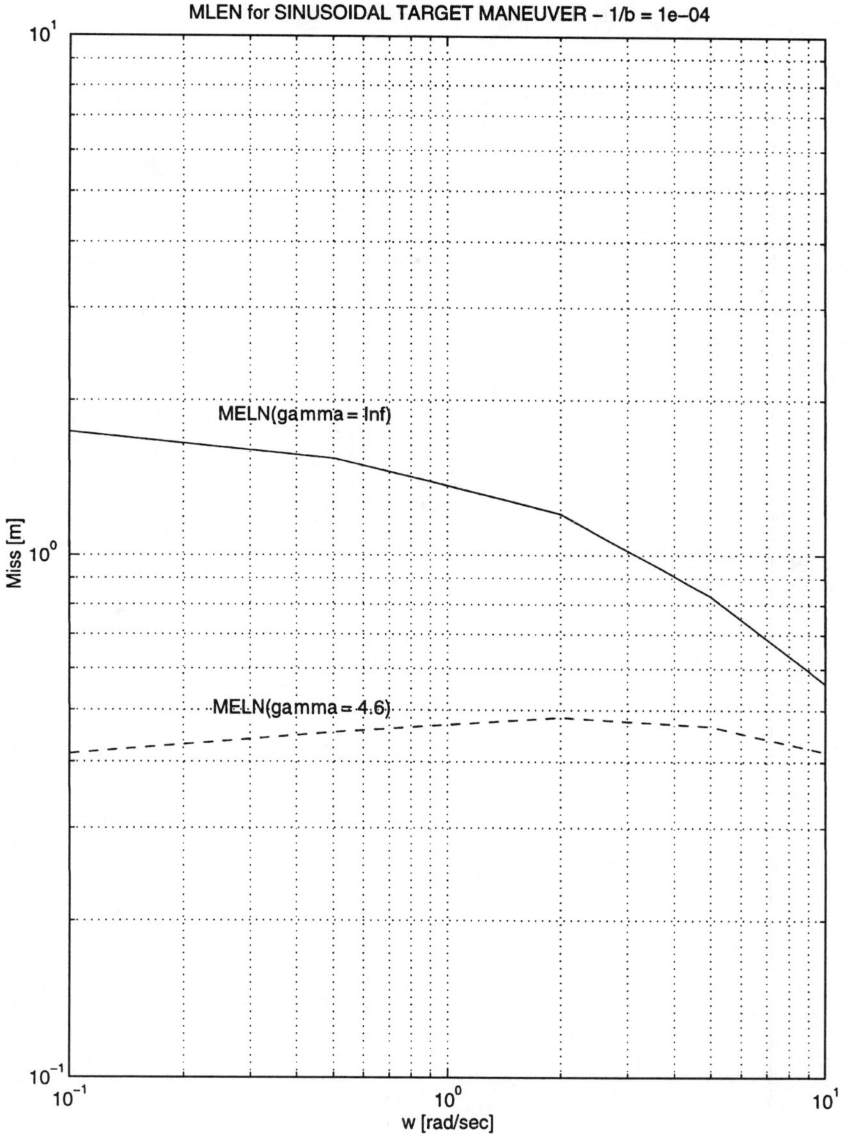

Fig. 4.12 Miss distance due to sinusoidal target maneuver.

MEL ($\gamma = 3.3$): This is the differential games based variant of the MEL [Eqs. (4.45)], where although the evader's time constant is ignored, its possibly time-varying (but of finite energy) maneuvers are considered. Here we also choose $1/b = 0.001$. We have chosen γ, which is slightly larger than the minimum γ.

SGL ($\gamma = 3.7$): This is the differential games based variant of SGL of Eqs. (4.27–4.29) that does not ignore either the pursuer's or the evader's time constants. It assumes time-varying finite-energy evader maneuvers. Here also we chose a nearly minimal γ.

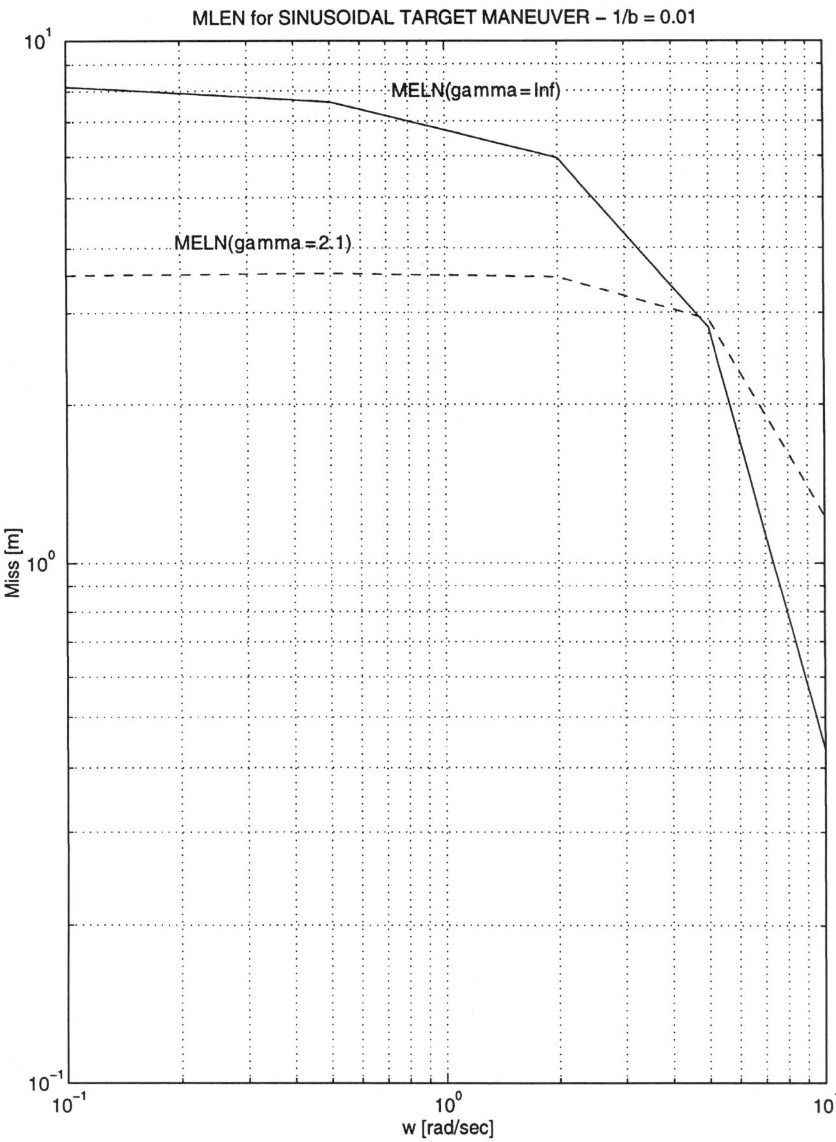

Fig. 4.13 Miss distance due to sinusoidal target maneuver.

The pursuit-evasion conflict we consider is characterized, as in the previous sections, by a closing velocity of 300 m/s of either head-on or tail-chase configuration and a pursuer time constant of 0.5 s. The evader's time constant is assumed to be 0.05 s. We compare the preceding four guidance laws for a sinusoidal target maneuver of 3 g of magnitude and 10 rad/s frequency. We assume that the pursuer commands are limited by 50 g.

The simulation results are given in Figs. 4.14–4.16. In Fig. 4.14 the relative separations that are obtained from the four guidance laws are shown. All guidance

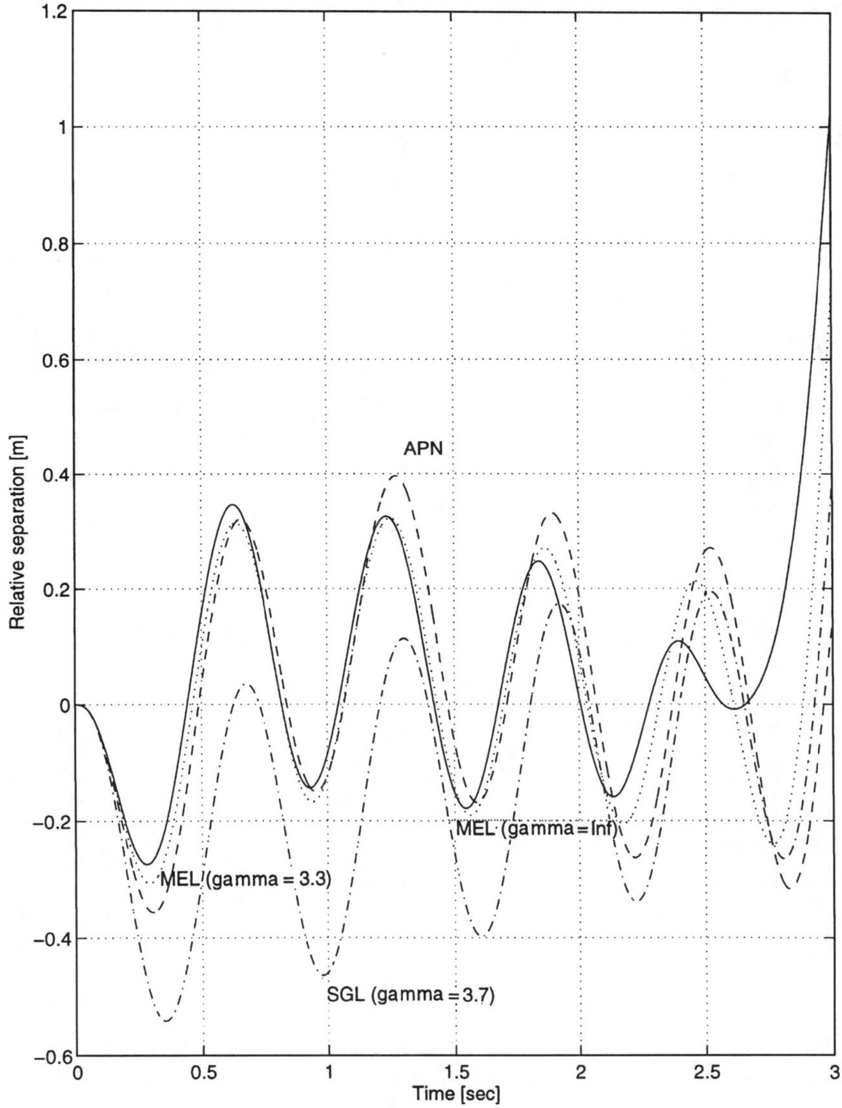

Fig. 4.14 Relative separation due to sinusoidal target maneuver.

laws other than the MEL ($\gamma \to \infty$) show similar behavior. The MEL ($\gamma \to \infty$) shows larger miss distance than the others. In Fig. 4.15 the commanded accelerations that are used by the pursuer are compared. We see that the APN makes the largest maneuvers (toward the conflict end) and also the MEL ($\gamma = 3.3$) applies rather large oscillatory maneuvers. The MEL ($\gamma \to \infty$) uses smaller accelerations than the MEL ($\gamma = 3.3$), which is an interesting result by itself. Both of the MEL variants ignored the evader's time constant, but the one with $\gamma = 3.3$ assumed also a possibly time-varying (worst-case) unmeasured evader maneuver. Since this did

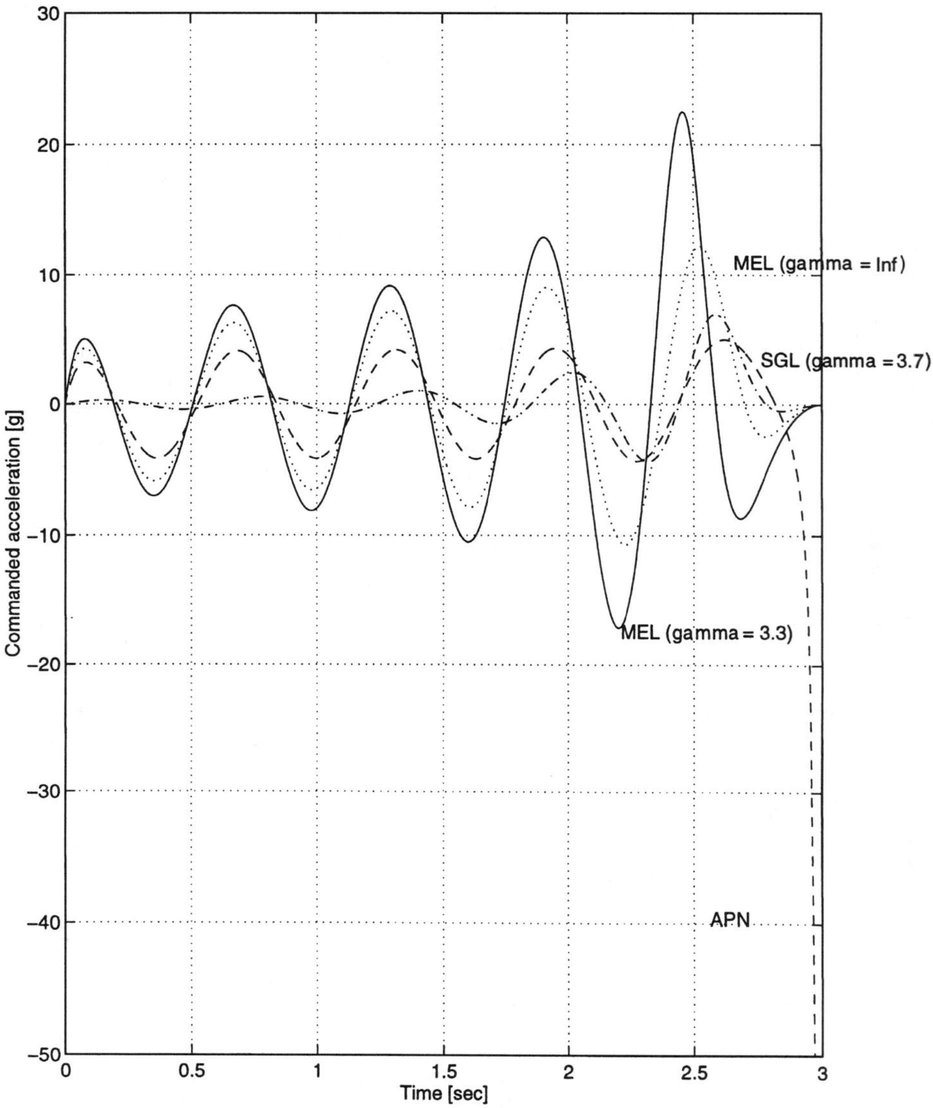

Fig. 4.15 Missile acceleration command due to sinusoidal target maneuver.

not happen (the whole maneuver was exactly measured), the MEL ($\gamma = 3.3$) operates under a situation that contradicts its design assumptions and, therefore, a poor performance is obtained. As could be expected, the SGL ($\gamma = 3.7$) uses the smallest pursuer maneuvers because it was designed exactly for the situation we are testing: quickly varying evader acceleration commands with a small evader time constant. The same results are seen in a clearer manner in Fig. 4.16 where the control effort is seen. Undoubtedly, $\gamma = 3.7$ is the one that is best tuned for the mission we tested.

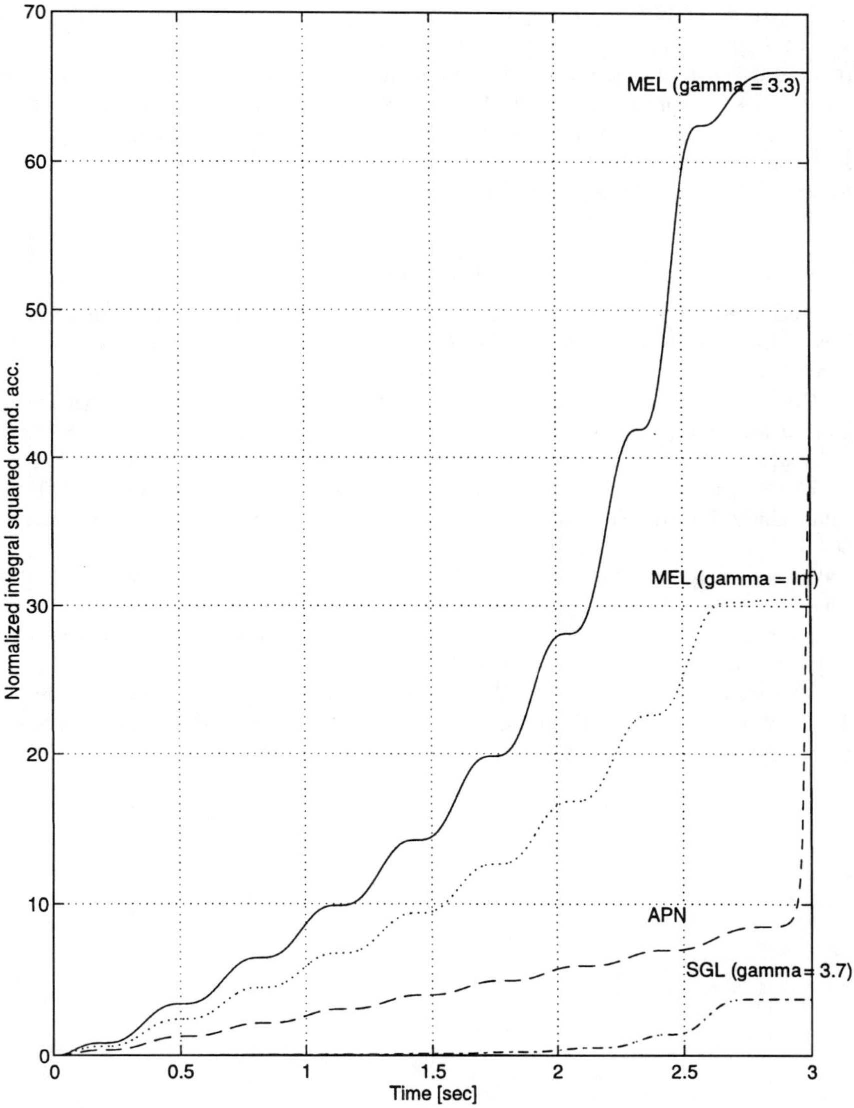

Fig. 4.16 Control effort due to sinusoidal target maneuver.

VII. Conclusions

In this chapter a new dimension was introduced into the derivation of various guidance laws by formulating differential games that correspond to different guidance problems. This new dimension enables us to model the evader's maneuverability against which we design our guidance laws. Together with the selection of a finite weighting coefficient of the miss distance we are capable of generating a continuum of guidance laws for different types of miss-distance causes (e.g., heading error, sinusoidal target maneuvers, steps). Avoiding conjugate points becomes a major design consideration under this formulation.

The numerical examples in this chapter emphasize the need in making good and realistic assumptions on the pursuer and the evader. The more realistic we make our assumptions, the better the performance of the resulting guidance law will be. Two assumptions that we have not yet doubted in this monograph are the exact knowledge of the time-to-go and also the exact knowledge of the pursuer and evader time constants. The sensitivity to errors in these assumptions and how to remedy this are discussed in Chapters 5 and 6.

References

[1]Anderson, G. M., "Comparison of Optimal Control and Differential Game Intercept Missile Guidance Laws," *Journal of Guidance and Control*, Vol. 4, No. 2, 1981, pp. 109–115.

[2]Ho, Y. C., Bryson, A. E., and Baron, S., "Differential-Games and Optimal Pursuit Evasion Strategies," *IEEE Transactions Auto. Control*, Vol. C-10, No. 4, Oct., 1965, p. 385.

[3]Isaacs, R., *Differential Games*, Wiley, New York, 1965, Chap. 1.

[4]Ben-Asher, J. Z., "Linear-Quadratic Pursuit-Evasion Games with Terminal Velocity Constraints," *Journal of Guidance, Control, and Dynamics*, Vol. 19, No. 2, 1996, pp. 499–501.

[5]Bryson, A. E., and Ho, Y. C., *Applied Optimal Control*, Hemisphere, New York, 1975, Chap. 9.

[6]Zarchan, P., *Tactical and Strategic Missile Guidance*, Vol. 124, Progress in Astronautics and Aeronautics, AIAA, Washington, DC, 1990, Chap. 7.

[7]Shaked, U., and Theodor, Y., "A Frequency Domain Approach to the Problems of H_∞ Minimum Error State Estimation and Deconvolution," *IEEE Transactions on Signal Processing*, Vol. 40, Dec., 1992, pp. 3001–3011.

Chapter 5

Optimal Guidance with Reduced Sensitivity to Time-to-Go Estimation Errors

I. Introduction

THE guidance laws obtained so far are strongly dependent on the missile's time constant and time-to-go. In real-life situations the knowledge of these two parameters involves inaccuracies. Considering the missile's time constant, two naturally raised practical questions are: 1) what happens if the missile's dynamics are not perfectly known, and 2) how does this uncertainty affect the performance of the guidance laws. These issues will be addressed in Chapter 4 where an optimal interception problem will be considered that takes into account the uncertainty of the time constants.

An equally (if not more) important issue is the effect of the time-to-go estimation error. It has been shown[1] that this estimation error can significantly degrade the performance of the optimal guidance law. The present chapter presents a straightforward approach for mitigating this effect.

The idea follows from the observation that the terminal relative lateral velocity $\dot{y}(t_f)$ also represents the sensitivity of the miss distance with respect to uncertainty in the time-to-go estimate. To show that this is so, we can use the result developed in previous chapters where we solved the interception problem with terminal velocity constraints, and where it was assumed that the guidance system had a negligible time lag. For the case of zero terminal conditions (game-theoretic optimal rendezvous) we have obtained [Eqs. (4.16)]

$$y(t) = c_1 t_{\text{go}}^2 + c_2 t_{\text{go}}^3 = c_1 (t_f - t)^2 + c_2 (t_f - t)^3$$

where $y(t)$ is the relative separation, t_f is the final time, and c_1 and c_2 are constants of integration. Notice that this differs from the game-theoretic optimal intercept where [Eqs. (4.12)]

$$y(t) = c_1 (t_f - t) + c_2 (t_f - t)^2$$

Recalling that $y(t_f)$ is the miss distance, we can evaluate the effect of small variation in t_f as follows:

$$y(t_f + \Delta t_f) = y(t_f) + \left.\frac{\partial y(t)}{\partial t_f}\right|_{t=t_f} \Delta t_f + O\left(\Delta t_f^2\right) \approx \left.\frac{\partial y(t)}{\partial t_f}\right|_{t=t_f} \Delta t_f$$

It is now clear why optimal rendezvous is less sensitive to uncertainty in t_f than optimal intercept because

$$\left.\frac{\partial y(t)}{\partial t_f}\right|_{t=t_f} = \begin{cases} 0 & \text{for rendezvous} \\ c_1 & \text{for intercept} \end{cases}$$

where, in general, $c_1 \neq 0$.

By employing either PN or APN with direct measurement of $\dot{\lambda}$, the sensitivity to time-to-go errors disappears. However, these strategies, as already mentioned, completely ignore the time constant T.

In this chapter (see also Ref. 2) the guidance system is a single lag. We formulate an LQ differential game problem with a cost that includes the adversaries' control efforts and miss distance, as well as a weighted combination of the terminal lateral velocity and the pursuer's acceleration. The latter is included for the purpose of relaxing the terminal acceleration requirements. The differential game approach rather than the one-sided optimization method is preferred because of its more general character and the fact that as $\gamma \to \infty$ nonmaneuvering target results are obtained. However, the ideas pursued henceforth (in this as well as the following chapters) are valid for both formulations.

For this case, direct integration of the state–costate equations turns out to be very difficult, and the problem will be solved by employing a Riccati equation formulation. A numerical parametric study will be performed employing various weights to demonstrate the robustness of the method.

II. Problem Formulation

We shall use the following assumptions from Chapter 1.

1) The evader directly controls its lateral acceleration while the pursuer is modeled as a first-order system.

2) The pursuer is more maneuverable than the evader.

3) The pursuer can measure its lateral acceleration in addition to the relative separation and velocity.

Under these assumptions the problem has the following state-space representation:

$$\dot{x} = Ax + Bu + Dw$$

$$A = \begin{bmatrix} 0 & 1 & 0 \\ 0 & 0 & 1 \\ 0 & 0 & -\dfrac{1}{T} \end{bmatrix} \qquad B = \begin{bmatrix} 0 \\ 0 \\ \dfrac{1}{T} \end{bmatrix} \qquad D = \begin{bmatrix} 0 \\ 1 \\ 0 \end{bmatrix} \tag{5.1}$$

where x_1, x_2, and x_3 (the components of x) are the relative separation, the lateral relative velocity, and the normal acceleration of the pursuer, respectively, and where u is the commanded acceleration of the pursuer and w is the normal acceleration of the evader (assumption 1).

The minmax problem is the following:

$$\min_u \max_w J = \frac{a}{2}x_1^2(t_f) + \frac{b}{2}x_2^2(t_f) + \frac{c}{2}x_3^2(t_f) + \frac{1}{2}\int_0^{t_f} u^2(t) - \gamma^2 w^2(t)\,\mathrm{dt} \tag{5.2}$$

The coefficient a weights the miss distance and is always positive. The coefficients b and c weight the terminal velocity and acceleration, respectively, and are nonnegative in this chapter. The former is used to obtain the desired sensitivity reduction to time-to-go errors, while the latter can be used to control the terminal acceleration. Evasive maneuvers are penalized by γ, which is required by assumption 2 to be greater than 1. Notice, finally, that Eqs. (4.22) present the solution for the problem in the special case where $b = c = 0$.

III. Optimal Feedback Solution

The Riccati equation for the general problem is the following (see Chapter 2):

$$-\dot{P} = PA + A^T P - PBB^T P + \gamma^{-2} PDD^T P \tag{5.3}$$

where the terminal conditions are

$$P(t_f) = \begin{bmatrix} a & 0 & 0 \\ 0 & b & 0 \\ 0 & 0 & c \end{bmatrix}$$

and the optimal strategies are of the form

$$\begin{aligned} u &= -B^T Px \equiv -(g_1 x_1 + g_2 x_2 + g_3 x_3) \\ w &= D^T \gamma^{-2} Px \end{aligned} \tag{5.4}$$

Let $S \equiv P^{-1}$ and define

$$S = \begin{bmatrix} S1 & S2 & S3 \\ S2 & S4 & S5 \\ S3 & S5 & S6 \end{bmatrix}$$

Thus

$$\dot{S} = AS + SA^T - BB^T + \gamma^{-2} DD^T \tag{5.5}$$

and

$$S(t_f) = \begin{bmatrix} a^{-1} & 0 & 0 \\ 0 & b^{-1} & 0 \\ 0 & 0 & c^{-1} \end{bmatrix}$$

Equation (5.5) can be rewritten explicitly as

$$\dot{S} = \begin{bmatrix} -2S2 & -S4 - S3 & -S5 + S3/T \\ & -2S5 - 1/\gamma^2 & -S6 + S5/T \\ & & 2S6/T + 1/T^2 \end{bmatrix} \tag{5.6}$$

Since the equation for $\dot{S}6$ is independent of the others, the solution for $S6$ can be easily obtained.

$$S6(t) = -\frac{1}{2T} + e^{2h}\left(\frac{1}{2T} + \frac{1}{c}\right) \tag{5.7}$$

Listing 5.1 Solving the Lyapunov equation

```
> # listing 5.1
> # Constructing the system matrices
> #
> with(linalg);
> A:=array([[0,1,0],[0,0,1],[0,0,−1/T]]);
> B:=array(1..3,1..1,[[0],[0],[1/T]]);
> D:=array(1..3,1..1,[[0],[1],[0]]);
> S(t):=array(1..3,1..3,[[S1(t),S2(t),S3(t)],[S2(t), S4(t),S5(t)],[S3(t),S5(t),S6(t)]]);
> #
> # Constructing the Lyapunov equation
> #
> temp:=evalm((A&*S(t)+S(t)&*transpose(A)));
> SD:=evalm((−temp+B&*transpose(B)−beta^(−2)*D&*transpose(D)));
> convert(SD,array);
> # solving for S6
> dsolve({diff(S6(t),t)=−SD[3,3],S6(Tf)=1/c},S6(t));
> assign(");
> # solving for S5
> dsolve({diff(S5(t),t)=−SD[3,2],S5(Tf)=0},S5(t));
> assign(");
> # solving for S4
> dsolve({diff(S4(t),t)=−SD[2,2],S4(Tf)=1/b},S4(t));
> assign(");
> # solving for S3
> dsolve({diff(S3(t),t)=−SD[1,3],S3(Tf)=0},S3(t));
> assign(");
> # solving for S2
> dsolve({diff(S2(t),t)=−SD[1,2],S2(Tf)=0},S2(t));
> assign(");
> # solving for S1
> dsolve({diff(S1(t),t)=−SD[1,1],S1(Tf)=1/a},S1(t));
> assign(");
```

where $h = \tau/T$ and $\tau = t_f - t$. We now can continue to solve sequentially for $S5, \ldots, S1$ (see Listing 5.1).

The following results have been obtained:

$$S5(t) = -\frac{1}{2} - \frac{1}{2}e^{2h} - \frac{Te^{2h}}{c} + \frac{e^h(c+T)}{c} \tag{5.8}$$

$$S4(t) = \tau + \frac{T}{2}e^{2h} + \frac{T^2}{c}e^{2h} - 2Te^h - \frac{2T^2}{c}e^h - \frac{\tau}{\gamma^2} + \frac{3T}{2} + \frac{T^2}{c} + \frac{1}{b} \tag{5.9}$$

$$S3(t) = -\frac{T}{2} + \frac{T}{2}e^{2h} + \frac{T^2e^{2h}}{c} - e^h\tau - \frac{e^hT\tau}{c} - \frac{T^2e^h}{c} \tag{5.10}$$

$$S2(t) = -\frac{\tau^2}{2} - \left(\frac{T^2}{2} + \frac{T^3}{c}\right)e^{2h} - \frac{T^2c + 2T^3}{2c} + \left(T^2 - \frac{2T^3}{c} + T\tau + \frac{T^2\tau}{c}\right)e^h + \frac{\tau^2}{2\gamma^2} - T\tau - \frac{T^2\tau}{c} - \frac{\tau}{b} \tag{5.11}$$

$$S1(t) = \frac{\tau^3}{3} + \left(\frac{T^3}{2} + \frac{T^4}{c}\right)e^{2h} - \left(2T^2\tau - \frac{2T^4}{c} + \frac{2T^3\tau}{c}\right)e^h - \frac{\tau^3}{3\gamma^2} + T\tau^2 + T^2\tau + \frac{T^4 + 2T^3\tau}{c} + \frac{T^2\tau^2}{c} + \frac{\tau^2}{b} + \frac{1}{a} \tag{5.12}$$

The feedback gains now can be obtained by inverting S and by using Eqs. (5.4). Because of the complexity of the results, this stage is best performed numerically rather than symbolically.

IV. MATLAB Simulation

To obtain numerical results, we used a MATLAB simulation program with the dynamics of Eqs. (5.1), similar to the program introduced in Chapter 3. However, since we do not have an analytic (closed-form) expression for the guidance law feedback gains as in the previous chapters, we have to calculate the Riccati matrix and the gains numerically. To this end, we calculate $S1$–$S6$ by Eqs. (5.7–5.12) and the state feedback gains are computed according to Eqs. (5.4) by

$$[g_1 \quad g_2 \quad g_3] = [0 \quad 0 \quad 1/T]\begin{bmatrix} S1 & S2 & S3 \\ S2 & S4 & S5 \\ S3 & S5 & S6 \end{bmatrix}^{-1} \tag{5.13}$$

Listing 5.2 contains a MATLAB listing of the feedback gain subroutine.

V. Numerical Results

A. Interception Conflict Description, Guidance Laws, and Error Sources

In this section, we consider a numerical example that illustrates the merits of the reduced sensitivity law (RSL). We analyze the effect on the miss distance of 3 g of constant acceleration target maneuver, 3 g of sinusoidal target maneuver of 2-rad/s frequency, and a 10-deg heading error. The conflict is assumed to take 3 s, and the time constant T of Eqs. (5.1) is taken to be 0.5 s. The conflict is assumed to be a head-on or tail chase or with a closing velocity of 300 m/s.

The effect of the following three guidance laws will be analyzed: PN, MEL (which includes a target acceleration feedback), and RSL. The design parameters for the PN guidance law is $N' = 3$, for the MEL, $\gamma = \infty$, and for the RSL, $a = 10000$, $b = 1000$, $c = 1000$, and $\gamma = 12$.

Remark: This value of γ is larger than the minimum γ, which still allows a bounded solution to Eq. (5.3) within $t \in [0, 3]$.

Notice that the LOS rate $\dot{\lambda}$ is assumed to be exactly measured and is not affected by possible errors in the time-to-go estimation. Therefore, unlike the MEL and

Listing 5.2 Feedback gains

```
function k = rsl_gains(t,T,a,b,c,gam)
%
% k = rsl_gains(t,T,a,b,c,gam)
%
% Robust guidance law with low sensitivity to time-to-go errors.
% Gains for :
%
% dx/dt = Ax + Bu + Dw
%
% A = [0 1 0      B = [0 0 1/T]'    D = [0 1 0]'
%      0 0 1
%      0 0 -1/T]
%
% J = x' (tf)*diag([a b c])*x(tf)^2 + ||u||^2 - gamma^2 * ||w||^2 ->saddle point
%
% t     : time-to-go
% T     : pursuer time constant
% gam   : evader maneuver weighting
h1 = exp(t/T);
h2 = exp(2*t/T);
t2 = t^2;
t3 = t2*t;
t4 = t3*t;
T2 = T^2;
T3 = T2*T;
T4 = T3*T;
gam2 = gam^2;
s1 = t3/3 + (T3/2+T4/c)*h2 -2*h1*T4/c - t3/(gam2*3) + T*t2 +...
T2*t2/c + t2/b -2*T2*h1*t - 2*T3*h1*t/c + T2*t + 2*T3*t/c - T3/2 ...
+ T4/c + 1/a;
s2 = -t2/2 - T2*h2/2 - T3*h2/c + h1*T2 ...
+ 2*h1*T3/c + t2/(2*gam2) ...
-T*t - T2*t/c - t/b + h1*T*t + T2*h1*t/c - (T2*c + 2*T3)/(2*c);
s3 = -T/2 + (T/2)*h2 + T2*h2/c - h1*t - h1*T*t/c - h1*T2/c;
s4 = t + (T/2)*h2 + T2*h2/c - 2*h1*T - 2*h1*T2/c - t/gam2 ...
+ 3*T/2 + T2/c + 1/b;
s5 = -1/2 - h2/2 - T*h2/c - h1*(-c-T)/c;
s6 = -1/(2*T) + h2*(1/(2*T)+1/c);
s = [s1 s2 s3;s2 s4 s5;s3 s5 s6];
p = inv(s);
B = [0 0 1/T]' ;
D = [0 1 0]' ;
k = -B' *p ;
k(4)= 0;
return
end
```

the RSL, the performance of the PN guidance law will not depend on these errors. Two possible types of errors can corrupt the time-to-go estimation. These are multiplicative (i.e., scale factor) and additive (i.e., bias or noise) errors.

In other words,

$$\tau_{\text{meas}} = \alpha\tau + \beta + n \tag{5.14}$$

where α, β, and n denote the scale factor, bias error, and noise, respectively. The analysis of the effect of the noise error is outside the scope of this book. In the following the results of a thorough analysis of the effect of bias errors β in the range of [0, 0.5] s will be described. In order not to burden the reader with too many numerical results, the effects of scale-factor errors will be described, so that an idea about the general trends is given. All of the preceding guidance laws will be considered, namely, the MEL, RSL, and PN. Notice that the MEL uses target acceleration measurements, while the RSL and PN do not. Therefore, to allow for a meaningful comparison with the RSL, we also consider the previously described variant of the MEL (MELN), which does not use target acceleration measurements.

The effect on the interceptor's performance of a time-to-go estimation error that varies between 0 to 0.5 s is computed for the four guidance laws and all three error sources (target maneuvers and heading error). Thus in the feedback gain calculation the measured time-to-go of Eq. (5.14) will be used where α is 1 and n is zero. The comparison between the three guidance laws is made by examining the miss distance and the maximum commanded interceptor acceleration due to target acceleration and heading error.

Notice that we have assumed that the interceptor's time constant T is not subject to uncertainties. The situation we analyze may, therefore, correspond to a missile that applies a gain-scheduled autopilot allowing only insignificant variations of the closed-loop transfer function that relates the acceleration command to the actual acceleration. We note that the PN guidance law was included in the comparison because it is well known and widely applied in practice. However, the PN guidance law exhibits a divergence some time before the conflict end. Therefore, it may use very large accelerations that become unbounded at t_f, possibly resulting in a very small miss distance that may not be achieved in practice due to maneuver limits of the pursuer. To make the comparison to the other three guidance laws (MEL, RSL, and MELN) more equitable, we chose to include a hard limiter of 15 g on the commanded target acceleration in our simulations. This limit is reasonable, considering the 3 g of maneuvers we take for the target (either of step or of sinusoidal type). Keep in mind, however, that the effect of this limiter is not included in the results of the adjoint simulations that will be discussed in the next section. The adjoint analysis is based on a linear model of the closed loop.

B. Simulation Results—Miss Distance and Maximum Accelerations

To allow for a clear comparison, we classify the simulation results according to their error sources (i.e., target acceleration step, sinusoidal target acceleration, and heading error). The most meaningful comparison is of the MELN, RSL, and PN guidance laws that do not use target acceleration measurements. The MEL, which uses target acceleration measurements, is included in the comparison for reference and to highlight the fact that acceleration measurements can improve performance

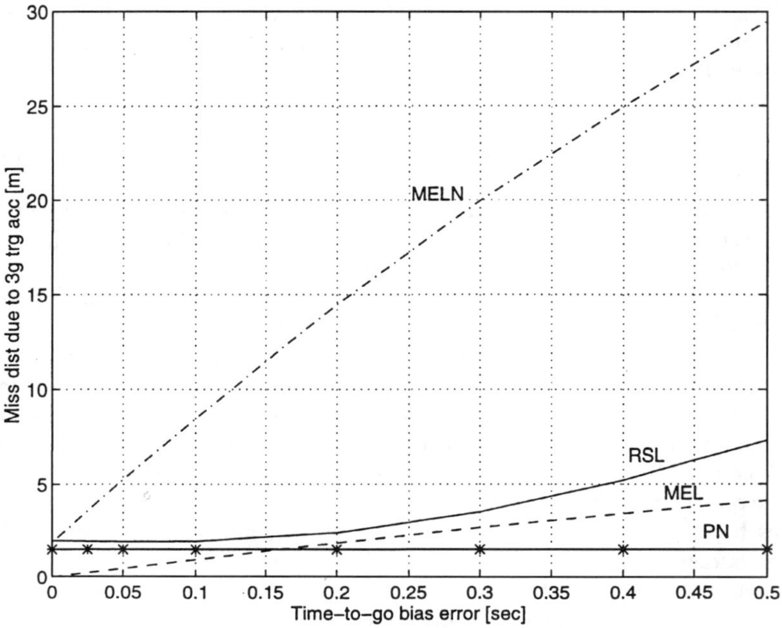

Fig. 5.1 Miss distance due to constant target maneuver.

when the target is maneuvering. The results are illustrated by comparing the graphs for the four guidance laws.

1. Step of 3 g *in Target Acceleration*

The RSL results in Fig. 5.1 obviously have a lower sensitivity to the time-to-go estimation error than MELN with considerably lower miss-distance values. PN is completely insensitive to the time-to-go estimation error and exhibits miss-distance values that are even smaller than those of the RSL. At first glance it may seem that if acceleration measurements are not available then the best choice is to use PN guidance law. This impression appears to be wrong if the maximum accelerations used throughout the interception conflict are compared (see Fig. 5.2). This comparison shows that the PN uses more than twice as much acceleration as RSL. MELN uses almost as large accelerations as PN but with a rather poor miss distance. Thus RSL seems to be the best choice if both miss distance and interceptor maneuvers are taken into account, and target acceleration measurements are not available. If, however, target acceleration measurements are available, the MEL has an advantage over RSL both in terms of miss distance and the maneuvers required. The nominal performance (i.e., zero error on the time-to-go estimate) is to be expected, since the MEL is especially tailored to constant target accelerations. What is surprising is its relative insensitivity to time-to-go estimation errors. The best choice when target acceleration measurements are available is, therefore, the MEL. However, if these measurements are not perfect (and they never are), the sensitivity to the measurement errors should be kept in mind and computed before finally deciding to use MEL. The preceding discussion does not change

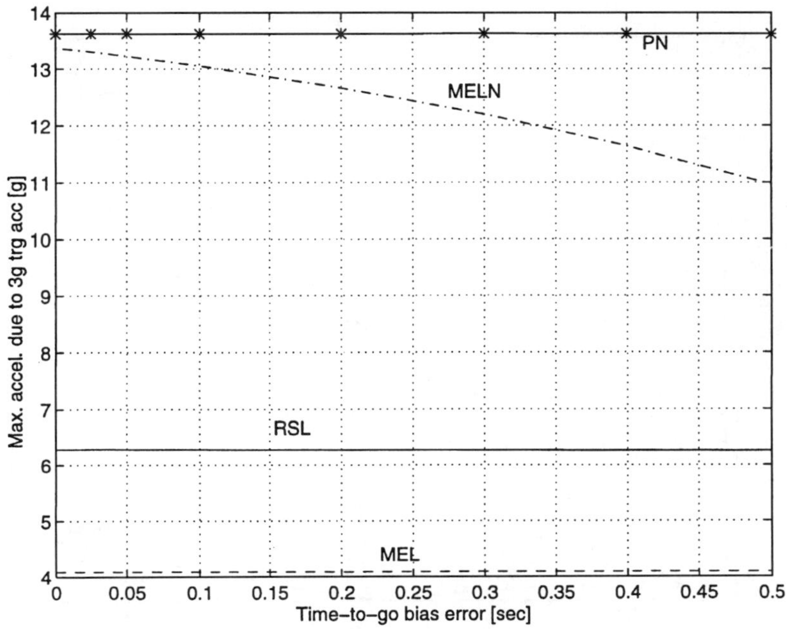

Fig. 5.2 Missile acceleration command due to constant target maneuver.

significantly as far as adjoint simulation results of Fig. 5.3 are concerned. These results were obtained for a fixed time-to-go error of 0.2 s, where the effect of acceleration limit was ignored.

2. *A 3* g*, 2 rad/s Sinusoidal Acceleration*

With this type of target maneuver, PN is no longer an adequate choice even for pursuers with unlimited maneuverability. It has poor miss-distance performance and requires more interceptor maneuvers than the RSL and MELN. Comparing the RSL and MELN performance shows (see Fig. 5.4) that for time-to-go errors that are less than about 0.1 s, the MELN has a somewhat lower miss distance than the RSL (about 0.5 m and 1 m, respectively) with the MELN requiring more maneuvers (8 g and 5 g, respectively; see Fig. 5.5). However, for time-to-go errors that are larger than about 0.2 s, the RSL provides lower miss distance at the cost of larger required maneuvers. This superiority is maintained for larger values of the error as well. Clearly, RSL is significantly less sensitive to time-to-go errors. Notice that MEL is also less sensitive than MELN to time-to-go estimation errors and is comparable to RSL (40% less acceleration required with somewhat inferior miss distances).

Remark: With sinusoidal maneuvers the frequency and the initial phase of the maneuver function affect the results. We are dealing with a single frequency only, and our conclusions concerning sinusoidal maneuvers should be taken with a great deal of caution. As previously discussed, we overcome the phase sensitivity of the results by taking phases within -180 deg to 180 deg with 45-deg steps, and the results shown are average values.

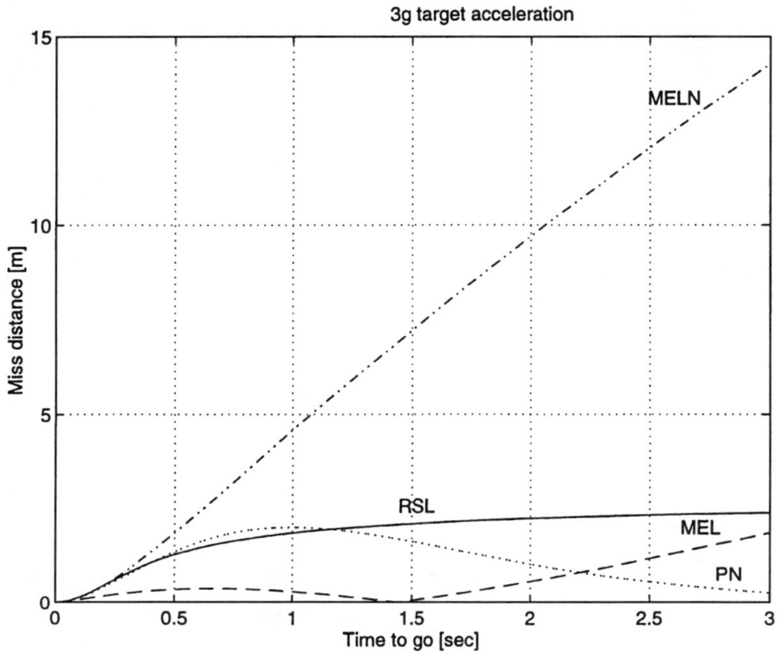

Fig. 5.3 Miss distance due to constant target maneuver and time-to-go error.

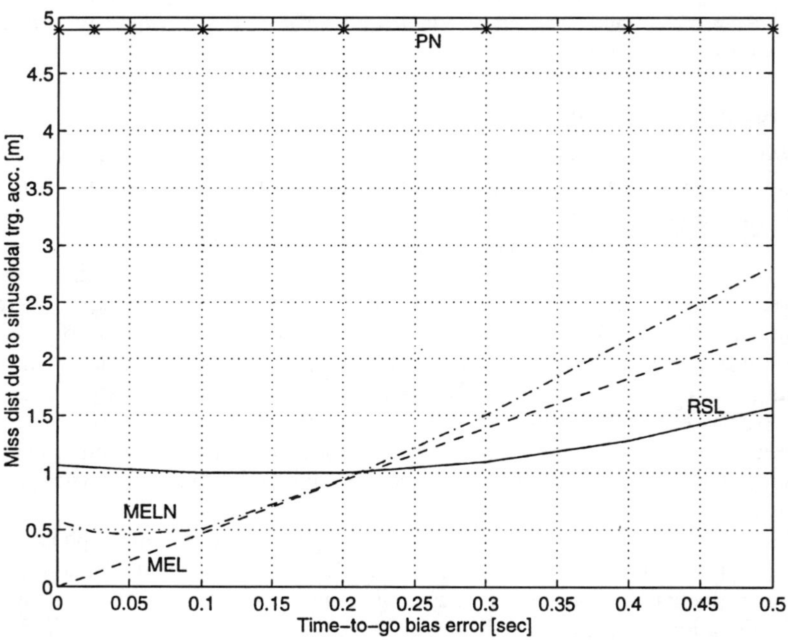

Fig. 5.4 Miss distance due to sinusoidal target maneuver.

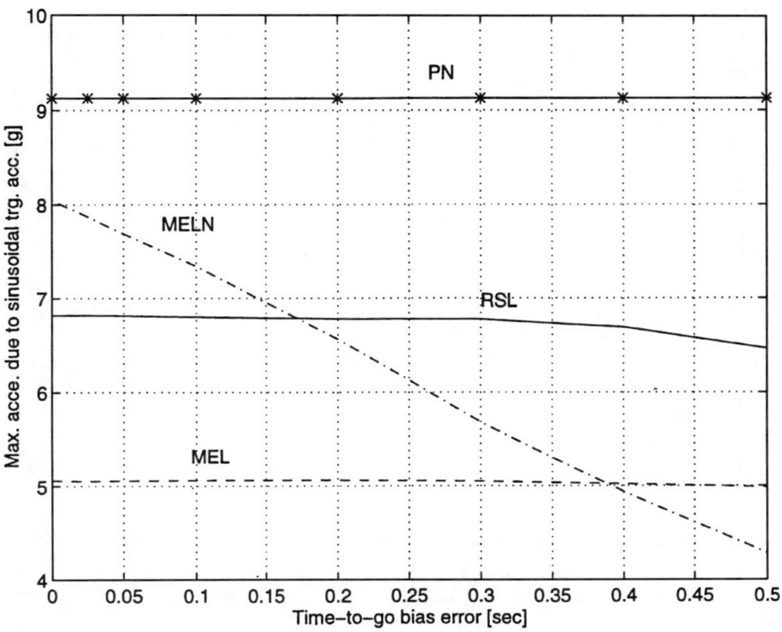

Fig. 5.5 Missile acceleration command due to sinusoidal target maneuver.

3. Heading Error of 10 Deg

For this case the RSL performance is the most impressive. The MELN, which coincides here with the MEL, has a miss distance that is almost linear with the time-to-go estimation error, with about 8-m miss distance for an estimation error of 0.2 s. In fact MEL is worse than PN for estimation errors larger than 0.03 s. The RSL, on the other hand, shows a very low sensitivity that leads to about 0.3-m miss distance for a time-to-go estimation error of 0.2 s (see Fig. 5.6) and less than 2 m with errors as large as 0.5 s. The required accelerations are still larger with the RSL (see Fig. 5.7). Also for short-duration interceptions (as short as about 1 s), the superiority of the RSL (in terms of miss distances) is maintained. This is shown by the adjoint simulations results for this case that are depicted in Fig. 5.8 (for a time-to-go error of 0.2 s).

Remark: Notice that we considered only positive bias errors on the time-to-go. Since at zero time-to-go, the MEL and MELN gains become singular, negative bias errors bring this singular point into the interception conflict time interval [see Eq. (5.14) and consider negative *b*]. Thus the commanded accelerations become unbounded. To overcome this difficulty, a number of policies can be adopted.

1) A fixed positive bias that is larger in its magnitude than the largest possible bias on Eq. (5.14) can be added. In such a case, the nominal performance will be degraded, but divergence will be prevented.

2) The guidance gains can be limited (i.e., above certain values they will be frozen).

3) Negative values of the estimated time-to-go can be avoided (i.e., set to zero if negative estimation is obtained).

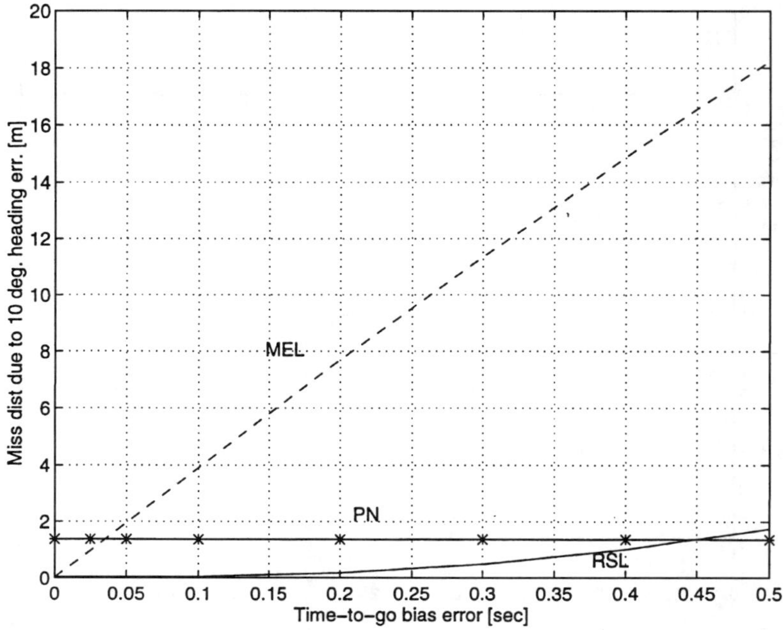

Fig. 5.6 Miss distance due to heading error.

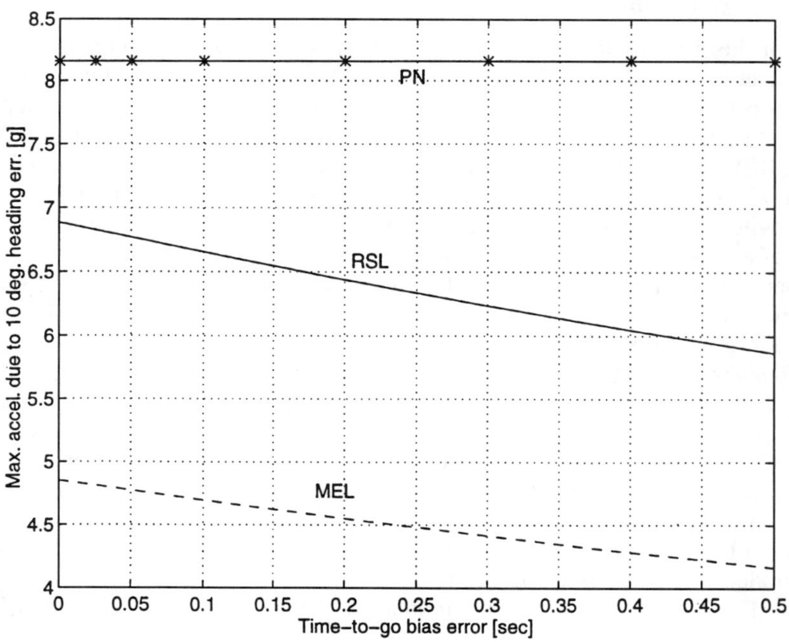

Fig. 5.7 Missile acceleration command due to heading error.

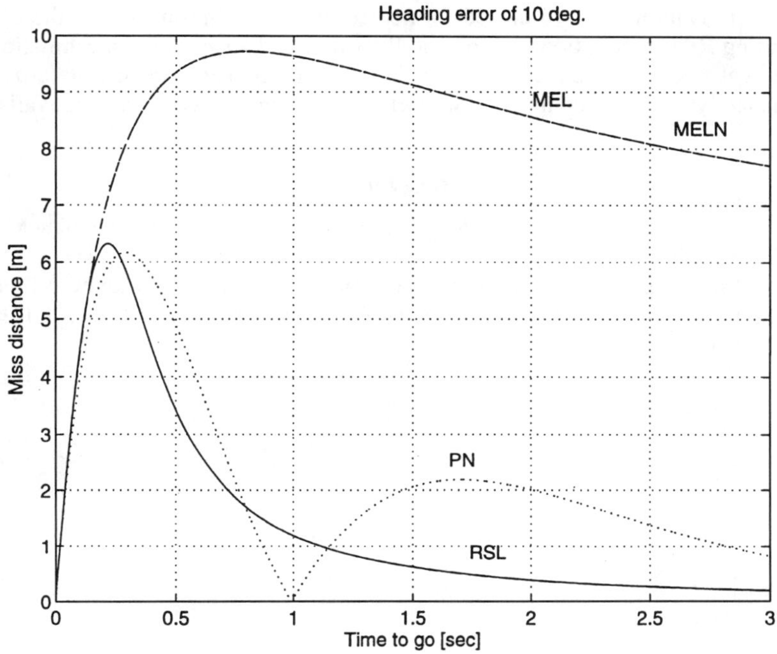

Fig. 5.8 Miss distance due to heading error and time-to-go error.

4) Freeze acceleration command above a certain limit.

The performance of all guidance laws (except for PN, which is insensitive) is obviously policy dependent for negative errors and will not be analyzed in this book.

Remark: As previously mentioned, another relevant time-to-go estimation error is scale factor. We also have conducted simulations for this type of error. For heading errors, the general trends are very much like those with bias errors. The RSL is clearly superior to MEL/MELN. It is also better than the PN (unlike the case of bias-type time-to-go errors). For constant acceleration maneuvering targets, the RSL still performs better than the MELN and worse than the MEL. For sinusoidal maneuvering targets, the results are indefinite with no clear advantage for any guidance law. To obtain a complete comparison, different frequencies should be checked. The numerical results for the effect of scale factors on the performance are not included.

VI. Conclusions

An optimal guidance law that reduces the effect of the time-to-go estimation error was developed. This guidance scheme is recommended for cases where target maneuvers are insignificant (e.g., stationary targets) or when the measurements of the target acceleration are not available. In such cases, the RSL outperforms the MEL and is better for reasonable time-to-go errors, even than the PN guidance law when mechanized via LOS rate measurement (i.e., inherently insensitive to time-to-go estimation errors).

The derivation leads to closed-formed formulas for the inverse of the corresponding Riccati equation. At present, the solution does not use an estimation of the target's acceleration. However, the idea of this chapter may be generalized to guidance systems that contain this important measurement whenever it is available.

References

[1]Nesline, W. N., and Zarchan P., "A New Look at Classical vs Modern Homing Missile Guidance," *Journal of Guidance and Control*, Vol. 4, No. 1, 1981, pp. 78–84.

[2]Ben-Asher, J. Z., and Yaesh, I., "Optimal Guidance with Reduced Sensitivity to Time-to-Go Estimation Errors," *Journal of Guidance, Control, and Dynamics*, Vol. 20, No. 1, 1997, pp. 158–163.

Chapter 6

Robust Guidance Methods

I. Introduction

THE methods presented so far have extensively used existing knowledge concerning various system parameters. In particular we assumed that the missile's time constant T is given. A practical question to ask is "What happens if this value is not perfectly known, and how does such an uncertainty affect the performance of the guidance laws?" (Recall that Chapter 5 dealt with the effect of the estimation error of the time-to-go.) The underlying assumption of the previously derived guidance laws concerning the constant speeds of the adversaries is valid along the pursuit phase, but the pursuer can have very different speeds in different conflicts depending on the time from launch to target acquisition and on the flight conditions. This can cause a significant uncertainty in the estimated values of T.

The main objective of this chapter is to explore a control policy obtained from a formulation of the optimal interception problem that takes into account the uncertainty of the pursuer's time constant. For the development we need conditions for optimality based on H_∞-optimal control theory that are not given in Chapter 2. We restrict the derivation here to specific results needed for our discussion without setting up the entire robust control theory that can be found in several references (e.g., Refs. 1 and 2). The main idea of this chapter was presented in Ref. 3 using a somewhat different problem formulation.

The chapter ends with a discussion concerning the issue of neglected dynamics. To this end we will analyze the performance of several guidance schemes for which the missile dynamics was assumed to be of first order for the design purposes while in fact a higher system represents the dynamics more adequately.

II. The Problem of Uncertain T

Assuming first-order lags for both the pursuer and the evader we have the following governing equation in state space:

$$\dot{x} = Ax + Bu + Dw$$

$$A = \begin{bmatrix} 0 & 1 & 0 & 0 \\ 0 & 0 & 1 & 1 \\ 0 & 0 & -\dfrac{1}{T} & 0 \\ 0 & 0 & 0 & -\dfrac{1}{\Theta} \end{bmatrix} \qquad B = \begin{bmatrix} 0 \\ 0 \\ \dfrac{1}{T} \\ 0 \end{bmatrix} \qquad D = \begin{bmatrix} 0 \\ 0 \\ 0 \\ \dfrac{1}{\Theta} \end{bmatrix} \tag{6.1}$$

The time constant T

$$T \equiv 1/a \tag{6.2}$$

is assumed to be in a prescribed interval, that is,

$$0 < a_0 \leq a \leq a_1 \tag{6.3}$$

so that the interceptor dynamics is open loop stable.

In the following formulation we let $x(0) = 0$, although the resulting guidance law is valid and will be demonstrated for other cases as well. We will also assume that $w(t)$ (the commanded acceleration of the evader) is of finite energy, that is,

$$\int_0^{t_f} w^2(t)\, \mathrm{d}t < \infty \tag{6.4}$$

Remark: The word *energy* is used loosely to indicate the control effort. This is common in texts dealing with H_∞-optimal control.

We define the following cost function:

$$J = x^T(t_f)Q_f x(t_f) + \int_0^{t_f} [u^2(t) - \gamma^2 w^2(t)]\, \mathrm{d}t \tag{6.5}$$

where

$$Q_f = \mathrm{diag}\{q, 0, 0, 0\} \qquad q > 0 \tag{6.6}$$

Our problem is to find a finite energy $u(t)$ that uses a full state feedback strategy, so that

$$J \leq 0 \tag{6.7}$$

for the minimum possible $\gamma = \gamma_0$ and for all a satisfying Eq. (6.3).

Remarks:

1) Minimization of γ, so that Eq. (6.7) is satisfied, corresponds to minimization of the effect of the target's maneuver on a weighted combination of the miss distance and the control effort. This fact can be better understood if Eqs. (6.5–6.7) are written for nontrivial $w(t)$ as:

$$\frac{qy^2(t_f) + \int_0^{t_f} u^2(t)\, \mathrm{d}t}{\int_0^{t_f} w^2(t)\, \mathrm{d}t} \leq \gamma^2 \rightarrow \min \tag{6.8}$$

2) Note that the disturbance attenuation problem of Eqs. (6.5–6.7) for $a_0 = a_1 = a$ is one part of the previously formulated LQ differential game (Chapter 4) where a saddle-point relation can be satisfied under certain conditions. In this game $w(t)$ maximizes J while $u(t)$ minimizes its value, so that

$$J(u^*, w) \leq J(u^*, w^*) \leq J(u, w^*) \tag{6.9}$$

This game is a zero-sum game [for $x(0) = 0$] where $J(u^*, w^*) = 0$ and thus the left inequality is equivalent to Eq. (6.7).

3) For $x(0) \neq 0$ the inequality (6.7) takes the form $J \leq J^*$, where J^* is positive and corresponds to the associated game.

The previously formulated problem has no closed-form solution for uncertain a in terms of a necessary and sufficient condition that ensures Eq. (6.7). We, therefore, use the method of Ref. 1, which gives only a sufficient condition and a rather simple solution for $u(t)$. This method applies an auxiliary disturbance attenuation problem that is defined (and related to our problem) in the next section.

III. The Auxiliary Problem

For convenience we make a small change in the notation

$$A = \begin{bmatrix} 0 & 1 & 0 & 0 \\ 0 & 0 & 1 & 1 \\ 0 & 0 & -a & 0 \\ 0 & 0 & 0 & -\dfrac{1}{\theta} \end{bmatrix} \qquad B_1 = \begin{bmatrix} 0 \\ 0 \\ 0 \\ \dfrac{1}{\theta} \end{bmatrix} \qquad B_2 = \begin{bmatrix} 0 \\ 0 \\ a \\ 0 \end{bmatrix} \tag{6.10}$$

The system of Eqs. (6.1) is then given by

$$\dot{x} = Ax + B_1 w + B_2 u \tag{6.11}$$

where A has an uncertain parameter. Together with the cost function of Eq. (6.7) our problem is a disturbance attenuation with an uncertain plant. Using the methods of Ref. 1, we can formulate an auxiliary problem for a completely known system where the uncertainty is represented by a corresponding finite-energy fictitious disturbance. To this end, we define

$$\delta \equiv (a_1 - a_0)/2 \tag{6.12}$$

and let Δ be a scalar with $|\Delta| \leq 1$ and also bring in a positive scaling factor ε. This scalar has no effect on the problem formulation, but it is going to affect the matrices L_a and R_a, which will be defined hereafter. The importance of this scaling idea will be discussed as follows.

We represent A as follows:

$$A = \begin{bmatrix} 0 & 1 & 0 & 0 \\ 0 & 0 & 1 & 1 \\ 0 & 0 & \alpha & 0 \\ 0 & 0 & 0 & -\dfrac{1}{\theta} \end{bmatrix} + \begin{bmatrix} 0 \\ 0 \\ \dfrac{1}{\varepsilon} \\ 0 \end{bmatrix} \Delta[0 \quad 0 \quad \delta\varepsilon \quad 0] \tag{6.13}$$

where

$$\alpha = -\frac{a_1 + a_0}{2} \tag{6.14}$$

We also represent B_2 by

$$B_2 = \begin{bmatrix} 0 \\ 0 \\ -\alpha \\ 0 \end{bmatrix} - \begin{bmatrix} 0 \\ 0 \\ \dfrac{1}{\varepsilon} \\ 0 \end{bmatrix} \Delta\delta\varepsilon \tag{6.15}$$

We denote

$$L_a = \begin{bmatrix} 0 \\ 0 \\ \frac{1}{\varepsilon} \\ 0 \end{bmatrix} \qquad R_a = [0 \quad 0 \quad \varepsilon\delta \quad 0] \qquad R_b = -\varepsilon\delta \tag{6.16}$$

and get the following representation of system (6.11)

$$\dot{x} = (A_\alpha + L_a \Delta R_a)x + B_1 w + (B_{2\alpha} + L_a \Delta R_b)u \tag{6.17}$$

where A_α and $B_{2\alpha}$ are the system matrices with the nominal value α.

We want to find a state feedback controller $u(t)$ that makes J in Eq. (6.5) nonpositive for all $|\Delta| \leq 1$. We use the following general result[1]:

Claim: Given the following system:

$$\dot{x} = (A_\alpha + L_a \Delta R_a)x + B_1 w + (B_{2\alpha} + L_a \Delta R_b)u \qquad x(0) = 0 \tag{6.18}$$

$$z = C_1 x + D_{12} u \tag{6.19}$$

and the cost function

$$J = x^T(t_f)Q_f x(t_f) + \int_0^{t_f} (z^T z - \gamma^2 w^T w)\, dt \tag{6.20}$$

then there exists a strategy for u that guarantees $J \leq 0$ for all $|\Delta| \leq 1$ and for all w s.t. $\int_0^{t_f} w^T w \, dt < \infty$ if the disturbance attenuation problem that is associated with the following system:

$$\dot{\tilde{x}} = A_\alpha \tilde{x} + \begin{bmatrix} L_a & \gamma^{-1} B_1 \end{bmatrix} \tilde{w} + B_{2\alpha} u \qquad \tilde{x}(0) = 0 \tag{6.21a}$$

$$\tilde{z} = \begin{bmatrix} R_a \\ C_1 \end{bmatrix} \tilde{x} + \begin{bmatrix} R_b \\ D_{12} \end{bmatrix} u \tag{6.21b}$$

is solvable, in the sense that there exists $u = G(t)\tilde{x}$ such that $J' \leq 0$ for all $\tilde{w}$ bounded by $\int_0^{t_f} \tilde{w}^T \tilde{w} \, dt < \infty$ where

$$J' = \tilde{x}^T(t_f)Q_f \tilde{x}(t_f) + \int_0^{t_f} (\tilde{z}^T \tilde{z} - \tilde{w}^T \tilde{w})\, dt \tag{6.22}$$

Moreover, the control that guarantees $J \leq 0$ is $u = G(t)x$.

Proof: This proof is somewhat heuristic, a more rigorous proof can be found in Ref. 1. Assume that $u = G(t)\tilde{x}$ is the solution for the accessory problem (this fact will be established later). From Eq. (6.22) we infer that the resultant cost satisfies

$$\tilde{x}^T(t_f)Q_f \tilde{x}(t_f) + \int_0^{t_f} (\tilde{z}^T \tilde{z})\, dt - \int_0^{t_f} (\tilde{w}^T \tilde{w})\, dt \leq 0 \tag{6.23a}$$

Define

$$y \equiv R_a \tilde{x} + R_b u \tag{6.23b}$$

and choose

$$\tilde{w} = \begin{bmatrix} \bar{w} \\ \gamma w \end{bmatrix} \quad \text{where} \quad \bar{w} \equiv \Delta y \quad \text{for} \quad |\Delta| \le 1 \tag{6.23c}$$

Notice that with this choice Eq. (6.18), the governing equation for x, and Eq. (6.21a), the governing equation for $\tilde{x}$, are identical (with the same control input u) and therefore $\tilde{x}(t) = x(t)$.

For the forthcoming arguments to be valid, the fact that $\int_0^{t_f} \tilde{w}^T \tilde{w} \, dt < \infty$ should be established. For a rigorous proof of this issue, the results of Ref. 2 can be applied. Consequently, we get that

$$(\tilde{z}^T \tilde{z}) = (y^T y) + (z^T z) \tag{6.23d}$$

$$(\tilde{w}^T \tilde{w}) = \gamma^2 (w^T w) + (\Delta y)^T (\Delta y) \tag{6.23e}$$

Substituting the last two expressions in Eq. (6.23a) we obtain (omitting the tilde from x in the quadratic terms)

$$x^T(t_f) Q_f x(t_f) + \int_0^{t_f} (y^T y + z^T z) \, dt - \int_0^{t_f} [\gamma^2 (w^T w) + (\Delta y)^T (\Delta y)] \, dt \le 0 \tag{6.23d$'$}$$

thus

$$x^T(t_f) Q_f x(t_f) + \int_0^{t_f} [z^T z - \gamma^2 (w^T w)] \, dt \le \int_0^{t_f} [(\Delta y)^T (\Delta y) - (y^T y)] \, dt \tag{6.23e$'$}$$

but

$$\int_0^{t_f} [(\Delta y)^T (\Delta y) - (y^T y)] \, dt \le 0 \qquad \text{for every} \quad |\Delta| \le 1 \tag{6.23f}$$

Hence we obtain that

$$x^T(t_f) Q_f x(t_f) + \int_0^{t_f} (z^T z - \gamma^2 w^T w) \, dt \le 0 \tag{6.23g}$$

for all $|\Delta| \le 1$. QED

For our problem we make now the following definitions:

$$\tilde{B}_1 = \begin{bmatrix} L_a & \gamma^{-1} B_1 \end{bmatrix} = \begin{bmatrix} 0 & 0 \\ 0 & 0 \\ \dfrac{1}{\varepsilon} & 0 \\ 0 & \dfrac{1}{(\gamma\theta)} \end{bmatrix} \tag{6.24}$$

$$\tilde{C}_1 = \begin{bmatrix} R_a \\ C_1 \end{bmatrix} = \begin{bmatrix} 0 & 0 & \delta\varepsilon & 0 \\ 0 & 0 & 0 & 0 \end{bmatrix} \tag{6.25}$$

$$\tilde{D}_{12} = \begin{bmatrix} R_b \\ D_{12} \end{bmatrix} = \begin{bmatrix} -\delta\varepsilon \\ 1 \end{bmatrix} \tag{6.26}$$

Thus

$$\dot{\tilde{x}} = A_\alpha \tilde{x} + \tilde{B}_1 \tilde{w} + B_{2\alpha} u \qquad \tilde{x}(0) = 0 \tag{6.27a}$$

$$\tilde{z} = \tilde{C}_1 \tilde{x} + \tilde{D}_{12} u \tag{6.27b}$$

We note that

$$R \equiv \tilde{D}_{12}^T \tilde{D}_{12} = [-\delta\varepsilon \quad 1][-\delta\varepsilon \quad 1]^T = 1 + \delta^2\varepsilon^2 > 0 \tag{6.28a}$$

and

$$\tilde{D}_{12}^T \tilde{C}_1 = [-\delta\varepsilon \quad 1] \begin{bmatrix} 0 & 0 & \delta\varepsilon & 0 \\ 0 & 0 & 0 & 0 \end{bmatrix} = [0 \quad 0 \quad \delta^2\varepsilon^2 \quad 0] \tag{6.28b}$$

The last expression yields a cross-product term of u and $\tilde{x}$ rendering the cost J nonstandard. The following claim provides the solution to this problem.

Claim: Consider the following matrix differential Riccati equation:

$$-\dot{P} = A_\alpha^T P + P A_\alpha - \left(P B_{2\alpha} + \tilde{C}_1^T \tilde{D}_{12}\right) R^{-1} \left(B_{2\alpha}^T P + \tilde{D}_{12}^T \tilde{C}_1\right) + P \tilde{B}_1 \tilde{B}_1^T P + \tilde{C}_1^T \tilde{C}_1 \tag{6.29}$$

$$P(t_f) = Q_f \tag{6.30}$$

Then $J' \leq 0$ if and only if there exists a solution $P(t)$ to this Riccati equation for all $t \in [0, t_f]$. Moreover, if such $P(t)$ exists, the solution for $u(t)$ is given by

$$u(t) = -R^{-1}\left(B_{2\alpha}^T P + \tilde{D}_{12}^T \tilde{C}_1\right)\tilde{x} \tag{6.31}$$

and the worst-case disturbance is given by

$$\tilde{w}(t) = R^{-1} \tilde{B}_1^T P \tilde{x} \tag{6.32}$$

Proof: A full derivation of necessary and sufficient conditions of this theorem in the style of Chapter 2 is beyond the scope of this book. Notice, however, the similarity with respect to the Riccati equations (2.24) and (2.25) where the main difference is the appearance of $\tilde{D}_{12}^T \tilde{C}_1$ in the quadratic term of the equation (and in the feedback gain). To convince the reader of the rationale of Eq. (6.29), the following change of variables (see Ref. 2) can be made:

$$\tilde{u}(t) = u + R^{-1} \tilde{D}_{12}^T \tilde{C}_1 \tilde{x} \tag{6.33a}$$

Then, the systems (6.27a) and (6.27b) become

$$\dot{\tilde{x}} = \bar{A} \tilde{x} + \tilde{B}_1 \tilde{w} + B_{2\alpha} \tilde{u} \qquad \tilde{x}(0) = 0 \tag{6.33b}$$

$$\tilde{z} = \bar{C}_1 \tilde{x} + \tilde{D}_{12} \tilde{u} \tag{6.33c}$$

where

$$\bar{A} = A_\alpha - B_{2\alpha} R^{-1} \tilde{D}_{12}^T \tilde{C}_1 \qquad \bar{C}_1 = \left(I - \tilde{D}_{12} R^{-1} \tilde{D}_{12}^T\right) \tilde{C}_1 \tag{6.33d}$$

Note that after this change of variables we get from Eq. (6.28a) that

$$\tilde{D}_{12}^T \bar{C}_1 = \tilde{D}_{12}^T (I - \tilde{D}_{12} R^{-1} \tilde{D}_{12}^T) \tilde{C}_1 = (\tilde{D}_{12}^T - \tilde{D}_{12}^T) \tilde{C}_1 = 0 \qquad (6.33e)$$

hence

$$\tilde{z}^T \tilde{z} = \tilde{x}^T (\bar{C}_1^T \bar{C}_1) \tilde{x} + \tilde{u}^T R \tilde{u} \qquad (6.33f)$$

Thus, by the results of Chapter 2, we obtain that the saddle-point control and disturbance for the auxiliary cost [Eq. (6.22)] are given by

$$\tilde{u}(t) = -R^{-1} B_{2\alpha}^T P \tilde{x} \qquad \tilde{w}(t) = R^{-1} \tilde{B}_1^T P \tilde{x} \qquad (6.33g)$$

where

$$-\dot{P} = \bar{A}^T P + P \bar{A} - (P B_{2\alpha}) R^{-1} (B_{2\alpha}^T P) + P \tilde{B}_1 \tilde{B}_1^T P + \bar{C}_1^T \bar{C}_1 \qquad (6.33h)$$

Noting that

$$\bar{C}_1^T \bar{C}_1 = \tilde{C}_1^T (I - \tilde{D}_{12} R^{-1} \tilde{D}_{12}^T) \tilde{C}_1 \geq 0 \qquad (6.33i)$$

and substituting $\bar{A} = A_\alpha - B_{2\alpha} R^{-1} \tilde{D}_{12}^T \tilde{C}_1$, we readily obtain Eq. (6.29) by arranging terms. Equation (6.31) is then obtained from Eqs. (6.33a) and (6.33g).

Remarks:

1) A saddle-point pair is formed by $u(t)$ and $\tilde{w}(t)$ for J' of Eq. (6.22). Moreover by the previous claim, the control strategy $u(t)$ ensures $J \leq 0$. It does not, however, mean that the original cost function J admits a saddle point.

2) Note that by varying ε we obtain a family of solutions rather than a single one, all of which satisfy the previously-formulated sufficiency condition for $J \leq 0$. A one-dimensional search can be carried out to select a favorable one.

IV. MATLAB Simulation

For the numerical results of this chapter, we used a MATLAB simulation program with the dynamics of Eqs. (6.1), similar to the program introduced in Chapter 3. However, since we do not have an analytic (closed-form) solution of the guidance law as in the previous chapters, we have to solve the Riccati differential equation (6.29) numerically. To this end, we change the free variable from t to $t_{go} = t_f - t$ and Eq. (6.29) is solved from $t_{go} = 0$ to $t_{go} = t_f$ by taking the initial condition $P(0) = Q_f$. The state feedback gains are then computed according to Eq. (6.31) by

$$K(t_{go}) = -R^{-1} [B_{2\alpha}^T P(t_{go}) + \tilde{D}_{12}^T \tilde{C}_1]$$

These gains are stored and used later in the forward simulation by reversing the arrays in which the gains are stored. Listing 6.1 shows of the feedback gains computation subroutine. The function *dif_ric_solve.m* (called by Listing 6.1) solves the Riccati equation numerically and is shown in Listing 6.2.

Listing 6.1 Feedback gains

```
%
% Compute gains of the robust guidance law (chap. 6) for the case of
% uncertain pursuer time constant
%
  THETA   = 0.5;             % evader time constant
  T_vec   = [0.5,1];         % pursuer time-constant region
  gam     = 4.2;             % disturbance-attenuation level
  one_over_b = 1e-03;        % miss-distance weighting
  epsilon1   = 1.5;          % relative-velocity weighting
  epsilon2   = 0.5;          % pursuer acceleration weighting
  epsilon    = 1.5;          % scaling parameter
  qf   = diag([1/one_over_b epsilon1 epsilon2 0]);
  % overall finite state weight
  % (see the remark at the end of section V.A)
  a0     = 1/T_vec(2);
  a1     = 1/T_vec(1);
  alfT   = -(a1 + a0)/2;
  delT   = (a1 - a0)/2;
%
% State-space description of the auxiliary game
%
  a   = [ 0 1 0   0
          0 0 1   1
          0 0 alfT 0
          0 0 0   -1/THETA];
  b1  = [0 0;0 0;1/epsilon 0;0 1/(gam*THETA)];
  b2  = [0 0 -alfT 0]';
  c1  = [0 0 delT*epsilon 0;0 0 0 0];
  d12 = [-delT*epsilon;1];
%
% gains computation
%
  tf  = 3;
  dt  = 0.001;
  k_vec = dif_ric_solve(a,b1,b2,c1,d12,qf,1,tf,dt);
```

V. Numerical Example

A. Interception Conflict Description, Guidance Laws, Uncertainties, and Design Parameters

To illustrate the merits of the robust guidance scheme, we consider a numerical example. We continue with the previously described conflict of 3-s duration and a closing velocity of 300 m/s. We analyze the effects on the miss distances of evasive target maneuvers, namely, a step in target acceleration of 3 g at the conflict start and an opposite maneuver of -3 g at time-to-go of τ s.

Listing 6.2 Solving the Riccati equation

```
function k=dif_ric_solve(a,b1,b2,c1,d12,qf,gam,tf,dt)
%
% Solves the diff. Riccati eq.:
%
% k=dif_ric_solve(a,b1,b2,c1,d12,qf,gam,tf,dt)
%
% -dp/dt = a'p + pa-(p*b2+c1'*d12)*r^ - 1(b2'*p+d12'*c1) +
%          p*b1*b1'* p/gam^2 + c1'*c1, p(tf)=qf
%
% and the corresponding gain : k = -r^-1(d12'*c1+b2'*p)
% is provided in reverse time;
%
t       = 0;
r       = d12'*d12;
d12c1   = d12'*c1;
c1c1    = c1'*c1;
ri      = inv(r);
p       = qf;
i       = 0;
p_dot = zeros(size(p));
k       = zeros(fix(tf/dt+1),length(a));
gami    = 1/gam;
for t=0:dt:tf,
   i          = i + 1;
   m2         = b2'*p + d12c1;
   m1         = b1'*p*gami;
   p_dot_prev = p_dot;
   p_dot      = a'*p + p*a-m2'*ri*m2 + m1'*m1 + c1c1;
   p          = p + (p_dot + p_dot_prev) * dt / 2;
   p          = (p + p') / 2;      % this is to keep p symmetric
                                   % in spite of numerical errors
   kx         = -ri*[d12c1+b2'*p];
   k(i,:)     = kx;
end
return
```

As indicated in Chapter 3 (Sec. V.C) this type of evasive maneuver is optimal under certain conditions. The following three guidance laws will be used for this numerical study: APN, MEL, and ROB (robust guidance law). All three guidance laws use a feedback from the evader's acceleration, which is assumed to be exactly known to the pursuer.

The design value for the airframe time constant is 0.5 s, but we assume that it can change up to 1 s. The evader's time constant is assumed to be 0.05 s. The evader is more agile than the pursuer and its time constant is much smaller than the conflict duration.

Remark: It is assumed that the airframe of the interceptor represents, in fact, the closed-loop transfer function of a normal acceleration control loop. Its time constant varies due to aerodynamic coefficient changes and operation point variations throughout the flight envelope.

The time-to-go τ where the evader changes its maneuver direction was chosen as 1 s since it had worst-case effects on all three guidance laws, especially in terms of the pursuer's control effort.

The design parameters for the APN guidance law is $N' = 3$; for the MEL, $1/b = 10^{-3}$ and $\gamma \to \infty$; and for the ROB guidance law, $1/b = 10^{-3}$, $\varepsilon = 1.5$, and $\gamma = 4.2$.

Remark: The latter value of γ is nearly the minimal value of the disturbance attenuation factor that was obtained with the specified ε. We also should note that for numerical reasons we took $Q_f = \text{diag}\{b, \varepsilon_1, \varepsilon_2, 0\}$, where $\varepsilon_1 = 1.5$ and $\varepsilon_2 = 0.5$.

B. Simulation Results

Simulations were performed for a couple of cases that will be referred to as nominal and perturbed for all three guidance laws.

Nominal: $T = 0.5$ s. This case for each guidance law will be called nominal although for the APN it has no implication since the APN does not assume knowledge of the pursuer's time constant.

Perturbed: $T = 1$ s. This case, for each guidance law, will be called perturbed with the same reservation as in nominal about the APN.

The resulting commanded accelerations are depicted in Fig. 6.1. We see that the APN uses large commanded pursuer accelerations with respect to MEL and ROB. We also see that the MEL is extremely sensitive to the pursuer's time constant, whereas the ROB shows a very moderate sensitivity as expected. This phenomena is best observed in Fig. 6.2 where the control effort for each guidance law and both cases (perturbed and nominal). The relative separations of the various conflicts we simulated are given in Fig. 6.3. We see that the ROB guidance law exhibits larger relative separation values than the MEL. This phenomenon can be explained because the ROB guidance law, being a differential game strategy, does not take for granted the assumption that the current evader's acceleration will maintain its value throughout the conflict. On the contrary, it is assumed that the target may change its maneuvers continuously, and hence the larger control activity is introduced toward the end of the conflict with the appropriate consideration of the time-lag uncertainty.

C. Higher Order Dynamics and Unstructured Uncertainties

It often happens that the dynamics of the pursuer is not only of first order. In fact, the order of the pursuer dynamics is the sum of the orders of the missile's open-loop dynamics (at least order 2 if short-period dynamics are taken,[4] plus the order of the compensation network of the acceleration control loop). In such cases, one may try to design the acceleration loop (if it is open to design or redesign), so that the dominant closed-loop transfer function of the acceleration control loop dynamics will be of first order. The pursuer dynamics will then be characterized by

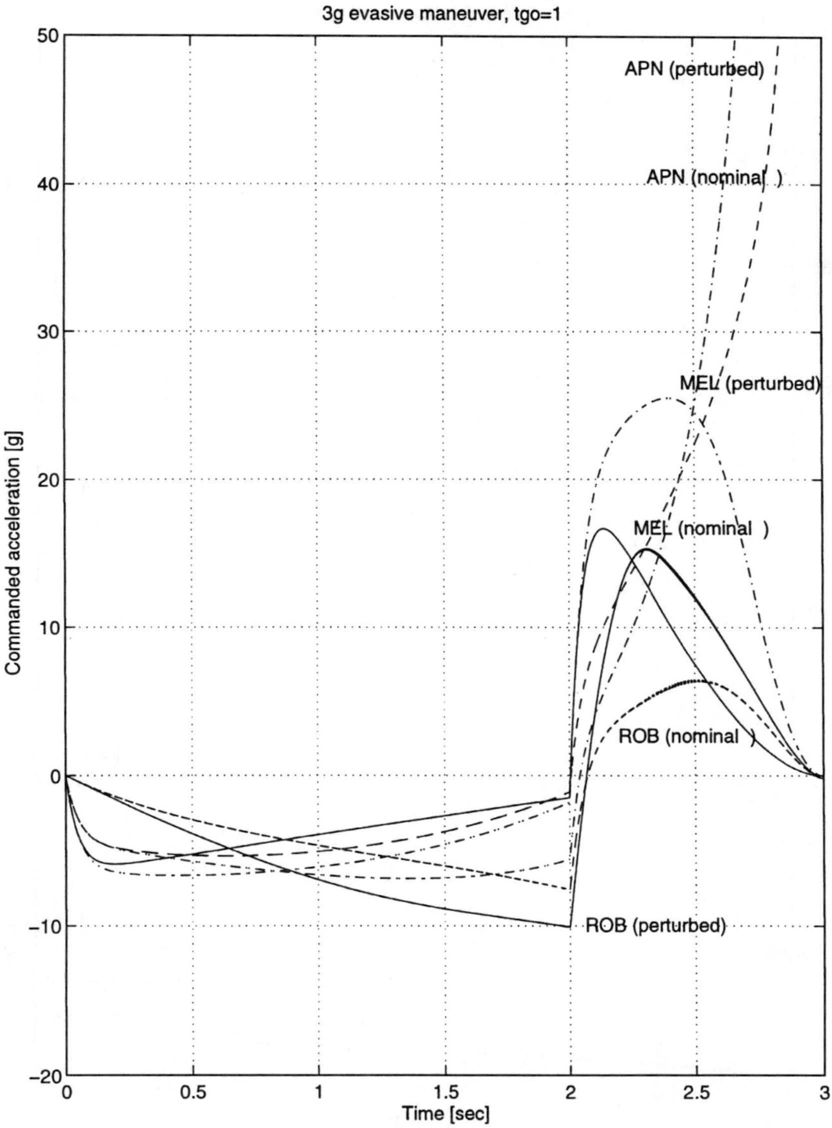

Fig. 6.1 Missile acceleration command due to bang-bang target maneuver.

a dominant (i.e., lowest real part of the pole) first-order lag, cascaded with higher order and of higher frequency (i.e., less dominant) dynamics.

Such a situation requires dealing with unstructured uncertainties of the dynamics,[2] which is defined by a frequency domain region of the closed loop's Bode plot. However, unstructured uncertainties are beyond the scope of this book, but we will examine how well the theory of this chapter for structured uncertainties copes with unstructured uncertainties. We will see via a numerical example that the ROB

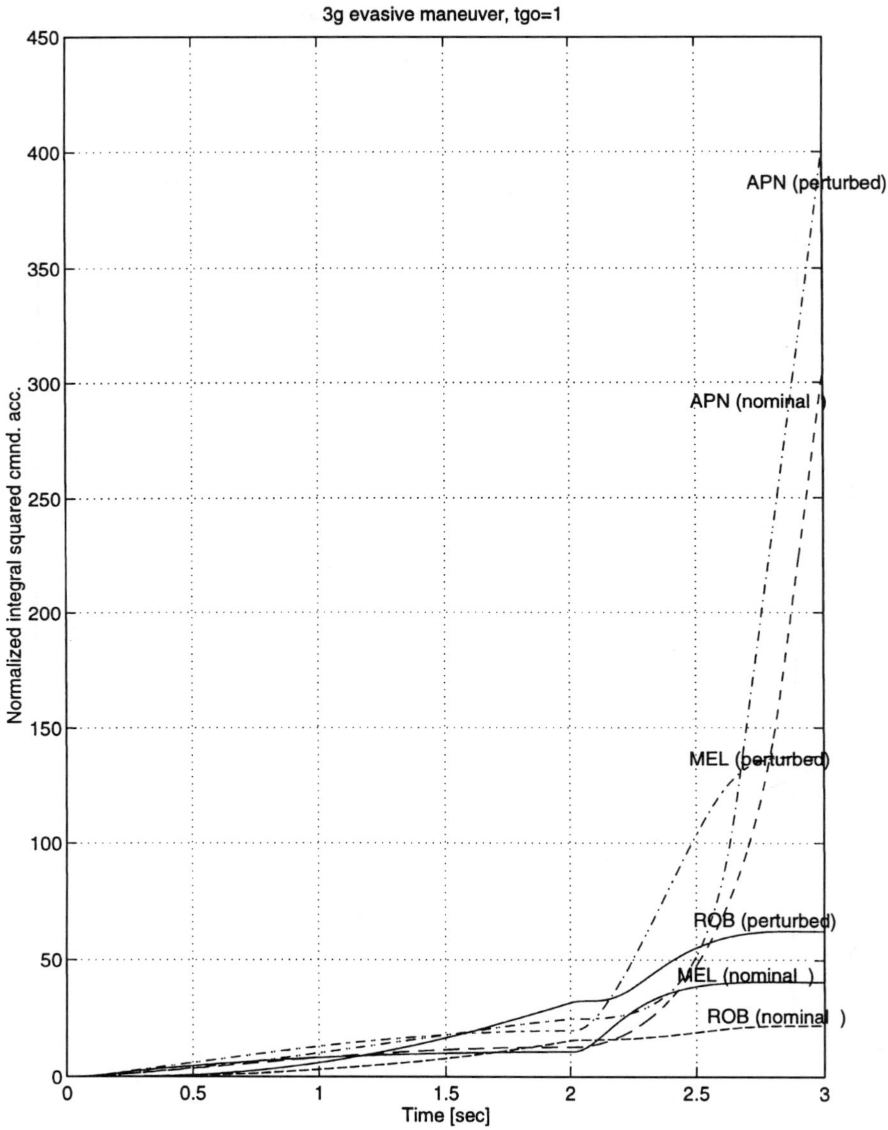

Fig. 6.2 Control effort due to bang-bang target maneuver.

guidance law provides some level of robustness also with respect to unstructured uncertainty. For the numerical study we have used the MATLAB program described in Appendix C (Sec. III) for general pursuer and evader dynamics.

We consider the same interception conflict parameters of Sec. V.A, and we assume the pursuer dynamics to be $1/(0.5s + 1)^2$. The long-term behavior (i.e., low-frequency characteristics) of this pursuer dynamics is expected to be similar to that of $1/(s + 1)$ and in fact is somewhere between $1/(s + 1)$ and $1/(0.5s + 1)$.

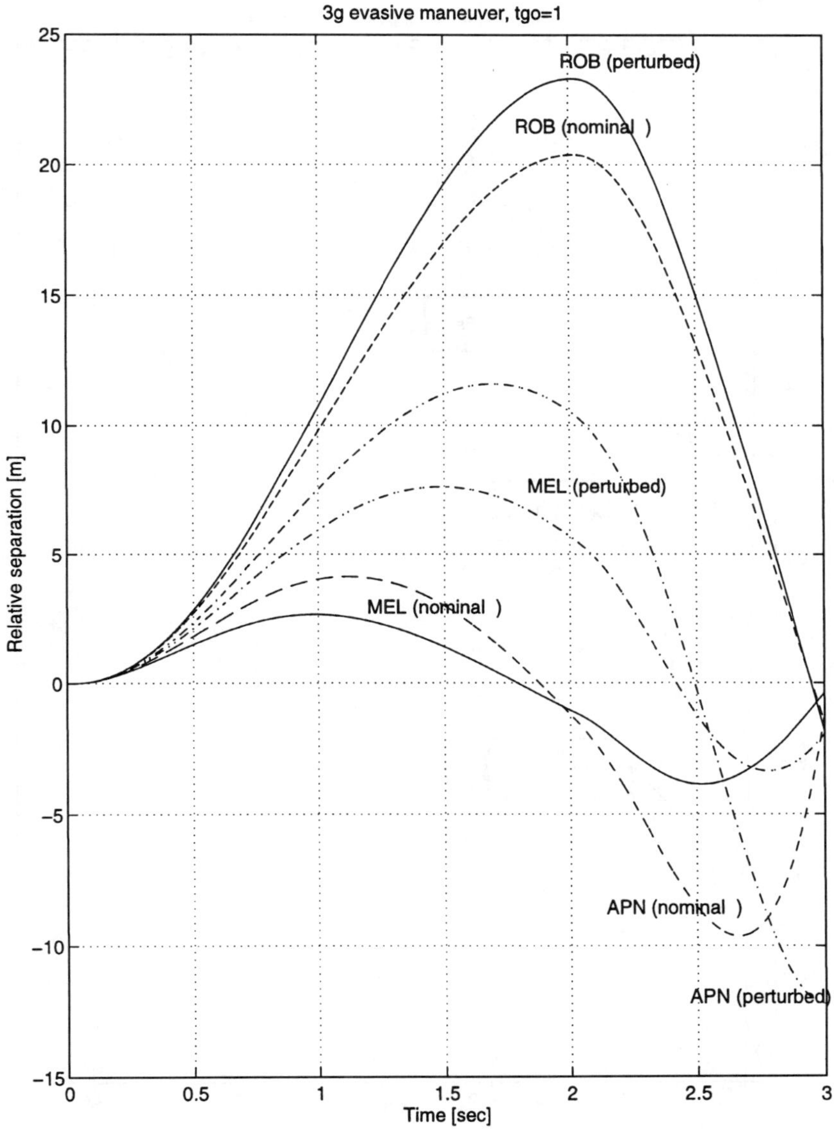

Fig. 6.3 Relative separation due to bang-bang target maneuver.

This can be verified by plotting the step responses of $1/(0.5s + 1)^2$ and those of $1/(s+1)$ and $1/(0.5s+1)$. The comparison in Fig. 6.4 of these three step responses justifies using the ROB guidance law for $T \in [0.5\ 1]$. The rest of the parameters are identical to those mentioned in Sec. V.A, and we compare two guidance laws: MEL, designed for $T = 0.5$, and ROB, designed for $T \in [0.5\ 1]$. We see in Fig. 6.5 the simulations of these two guidance laws for a 3 g step of target acceleration. The comparison of Fig. 6.5 indicates a lower miss distance for the ROB guidance law

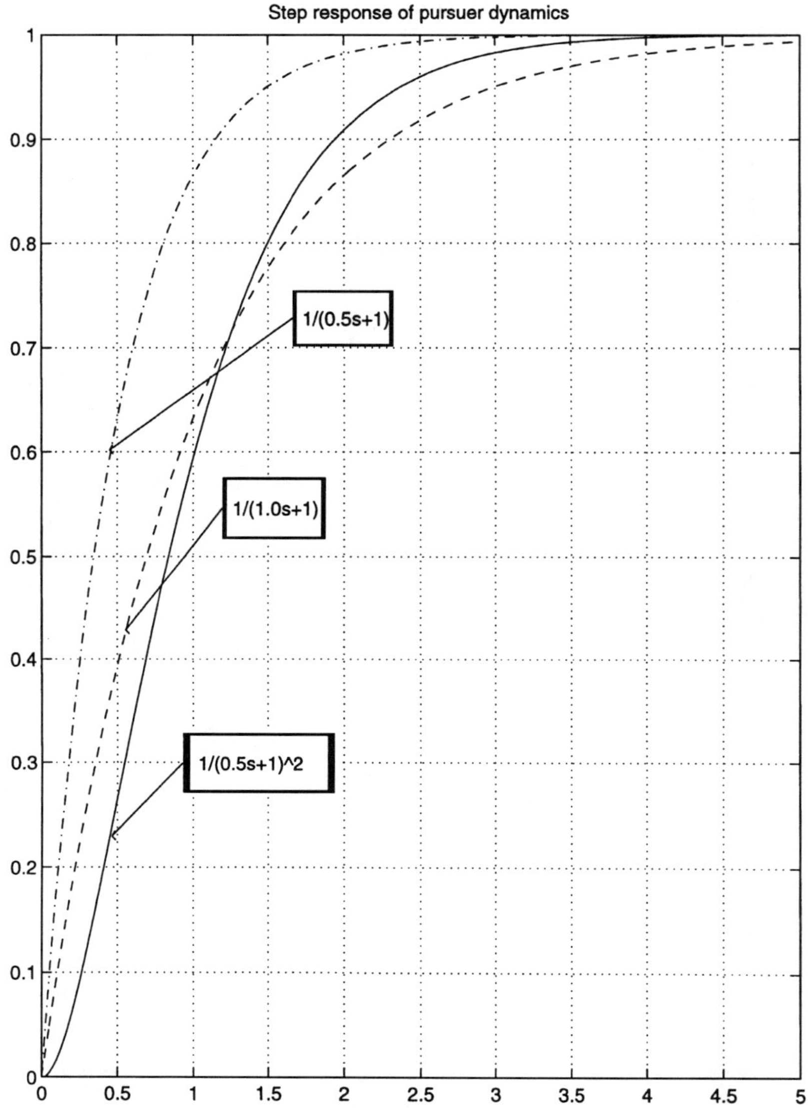

Fig. 6.4 Different models for missile dynamics.

with respect to the MEL. We also see from the forward simulations that the ROB guidance law requires significantly less control effort (see Fig. 6.6). We note that in cases where the pursuer dynamics do not correspond to somewhere in between the extreme values of T, the ROB guidance law is not expected to give good results.

VI. Conclusions

A robust guidance law that takes into account the uncertainty in the single time constant of the airframe and target evasive maneuvers was developed. The

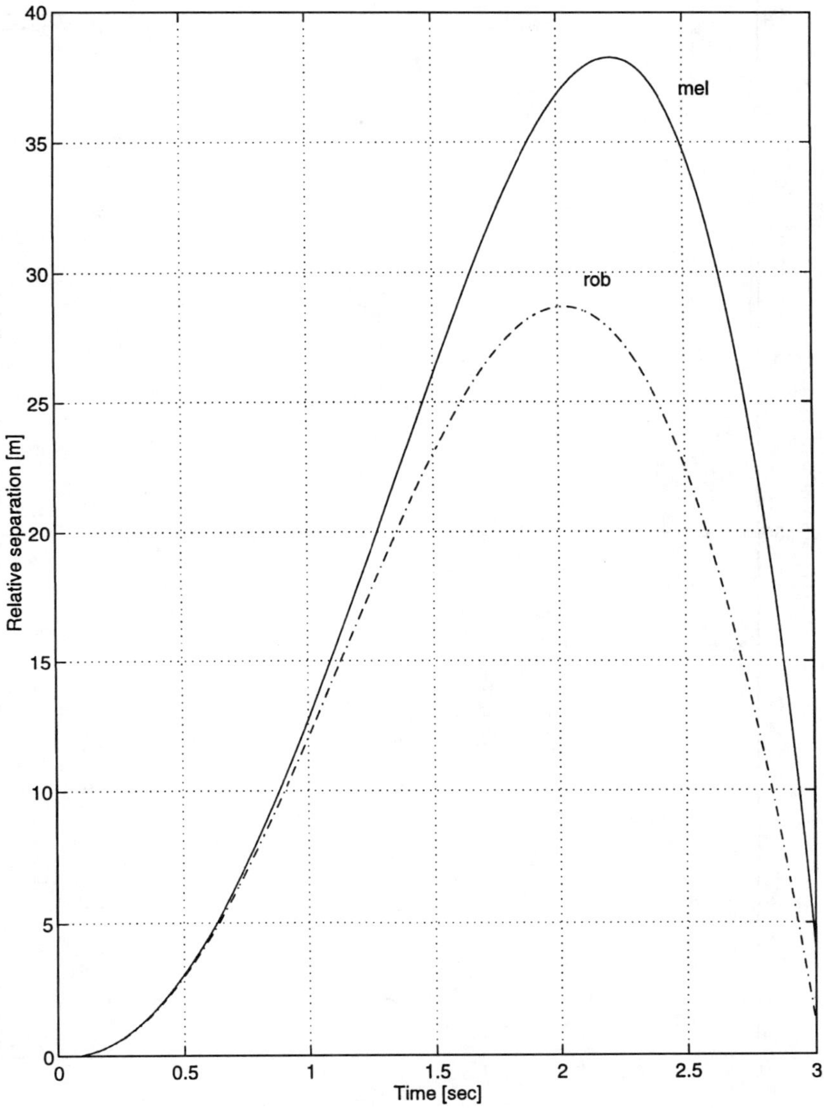

Fig. 6.5 Relative separation due to constant target maneuver.

derivation applied an H_∞-based theory of finite-time horizon disturbance attenuation in the presence of real-bounded time-varying uncertainties.

The solution obtained is sufficient but not necessary; hence, a better guidance law may exist that our approach has not revealed. The results of this chapter showed that some price has to be paid to achieve robustness. The penalty we pay is twofold: the performance when no uncertainty exists is somewhat degraded and the realization of the guidance law becomes more difficult (no closed-form solutions exist for the Riccati equation and the feedback is not simply related to the LOS rate).

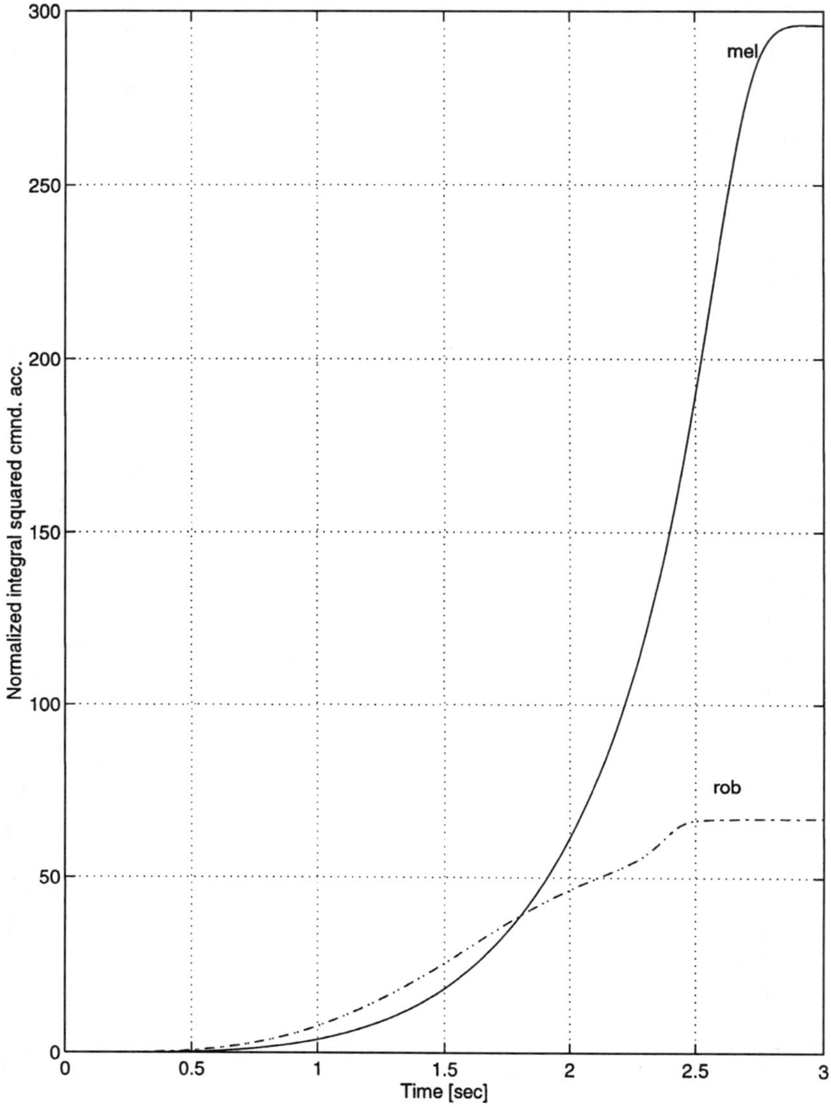

Fig. 6.6 Control effort due to constant target maneuver.

The guidance law was tested using forward simulations for nominal and perturbed interceptor dynamics and for some suboptimal evasive maneuvers. The simulation results are very promising and should encourage further study of the ROB guidance law, which showed an obvious superiority over the other tested guidance laws (namely, APN and MEL). The ROB guidance law also was tested and performed well for a case of more general unstructured uncertainties.

References

[1]de Souza, C. E., Shaked, U., and Fu, M., "Robust H_∞ Tracking: A Game Theory Approach," *International Journal of Robust and Nonlinear Control*, Vol. 5, No. 3, 1995, pp. 223–238.

[2]Green, M., and Limebeer D. J. N., *Linear Robust Control*, Prentice-Hall, Englewood Cliffs, NJ, 1995, Chap. 6.

[3]Yaesh, I., and Ben-Asher, J. Z., "Optimum Guidance with a Single Uncertain Time Lag," *Journal of Guidance, Control, and Dynamics*, Vol. 18, No. 5, 1995, pp. 981–988.

[4]Etkin, B., *Dynamics of Atmospheric Flight*, Wiley, New York, 1972, Chap. 9.

Chapter 7

Optimal Guidance with Multiple Targets

I. Introduction

THE guidance laws obtained so far have solved the optimal interception with a single target. The success of the previous methods provides the motivation for studying this problem with more than one target. Norbutas[1] addressed this problem for m targets and formulated a corresponding set of m two-point boundary-value problems. For the general case he suggested some numerical solution techniques for open-loop control, whereas for the two-target case he formulated a closed-form solution. This result was generalized by Ben-Asher[2] who solved the m-target case.

For the development we need results from optimal control theory concerning terminal and interior point constraints that are not given in Chapter 2. We shall present these results without giving the detailed derivation that can be found in many references (i.e., Ref. 3). The problem of intercepting m stationary targets will then be formulated and the solution will be obtained in closed form by employing cubic-spline interpolation.

II. Terminal and Interior Point Constraints

Consider the following optimal control problem

$$\min_u J = \frac{1}{2}\int_{t_0}^{t_f} u^2(t)\,\mathrm{d}t \tag{7.1}$$

subject to the linear state equation

$$\dot{x} = A\boldsymbol{x} + Bu \tag{7.2}$$

where $\boldsymbol{x}$ is an n-dimensional state vector and u is a scalar input signal, and to the following point constraints:

$$N_i[x(t_i)] = 0 \tag{7.3}$$

where N_i is a differentiable scalar function for $i = 1, \ldots, m$, and where $t_0 < t_i \le t_m = t_f$. Hence m is the number of constraints and N_m is the terminal one.

The optimal control for this problem is[3]

$$u = -B^T\boldsymbol{\lambda} \tag{7.4}$$

where the adjoint vector $\boldsymbol{\lambda}$ satisfies

$$\dot{\lambda} = -A^T\lambda \tag{7.5}$$

for $t \neq t_i$ but is, in general, discontinuous at $t_i < t_f$, such that

$$\lambda(t_i+) = \lambda(t_i-) + \nu_i \frac{\partial N_i}{\partial x} \tag{7.6}$$

where normally ν_i is a nonzero coefficient.

Finally, the costate vector should satisfy the transversality condition

$$\lambda(t_f) = \nu_m \frac{\partial N_m}{\partial x} \tag{7.7}$$

The proof will not be given here and can be found in many reference books on optimal control.[3]

III. Problem Formulation

Assume m stationary targets located in the plane in the neighborhood of a nominal LOS at fixed normal displacements $y_1, y_2, \ldots, y_m$. Without any loss of generality we let $y_m = 0$ (i.e., the nominal LOS is constructed with respect to the last target). The nominal times at which the pursuer approaches these targets (i.e., passes through their projections on the LOS) are $t_1, t_2, \ldots, t_m = t_f$.

Assuming an ideal pursuer, we get [see Eqs. (1.21)]

$$\dot{x} = x_2 \qquad \dot{x} = u \tag{7.8}$$

We consider the following optimal guidance problem

$$\min_u J = \frac{1}{2} \int_{t_0}^{t_f} u^2(t)\, \mathrm{d}t \tag{7.9}$$

subject to the following point constraints

$$x_1(t_i) = y_i \qquad \text{for } i = 1, 2, \ldots, m \tag{7.10}$$

Notice that the last constraint (prescribed at the terminal time) replaces the penalty term used so far in the cost function. The other $m - 1$ constraints are classified as interior point constraints. Notice also that $x_2(t)$ is unconstrained. Because of the structure of the cost, we may call the resulting guidance law minimum effort guidance (MEG).

IV. Optimal Solution

The optimal control problem is constrained by

$$N_i[x(t_i)] = x_1(t_i) - y_i \qquad \text{for } i = 1, 2, \ldots, m \tag{7.11}$$

hence

$$\frac{\partial N_i}{\partial x} = \begin{bmatrix} 1 \\ 0 \end{bmatrix} \qquad \text{for } i = 1, 2, \ldots, m \tag{7.12}$$

From Eqs. (7.4–7.7), we obtain by substituting

$$A = \begin{bmatrix} 0 & 1 \\ 0 & 0 \end{bmatrix} \qquad B = \begin{bmatrix} 0 \\ 1 \end{bmatrix}$$

that

$$u = -\lambda_2 \tag{7.13}$$

where the costate satisfies

$$\dot{\lambda}_1 = 0 \qquad \dot{\lambda}_2 = \lambda_1 \tag{7.14}$$

with the additional allowable $m - 1$ "jumps"

$$\begin{aligned} \lambda_1(t_i+) &= \lambda_1(t_i-) + \nu_i \\ \lambda_2(t_i+) &= \lambda_2(t_i-) \end{aligned} \tag{7.15}$$

for $i = 1, 2, \ldots, m - 1$.

Finally, at the terminal time ($t_f = t_m$) we require that

$$\lambda_1(t_m) = \nu_m \qquad \lambda_2(t_i+) = 0 \tag{7.16}$$

We make the following observations:

1) λ_2 and therefore u are continuous.

Remark: The strategy of sequentially intercepting target by target (solving separate problems) yields piecewise continuous control.

2) λ_1 is piecewise constant, and therefore λ_2 and u are piecewise linear.

3) λ_2 and therefore u vanish at the final time.

Since u is piecewise linear, x_2 is piecewise quadratic, and x_2 piecewise cubic (integrating twice the piecewise linear control).

The problem can now be transformed into one of cubic-spline interpolation. Our aim is to interpolate the points $[t_0, y_0 \equiv x_1(t_0)]$, (t_1, y_1), (t_2, y_2), $\ldots$, (t_m, y_m) with cubic polynomials while requiring the continuity of the first and second derivatives and satisfying a given first derivative $x_2(t_0)$ at t_0 and a zero second derivative at t_f.

This is an almost standard interpolation problem.[4] Let $A_0, A_1, \ldots, A_m$ denote the value of the control u at $t_0, t_1, \ldots, t_m$ and define $P_i''(t)$ to be a linear interpolation between A_{i-1} and A_i on the interval $[t_{i-1}, t_i]$. Thus

$$u(t) \equiv P_i''(t) = \frac{t_i - t}{h_i}A_{i-1} + \frac{t - t_{i-1}}{h_i}A_i \qquad \text{for } t \in [t_{i-1}, t_i] \quad i = 1, 2, \ldots, m \tag{7.17}$$

where $h_i = t_i - t_{i-1}$.

In addition we obtain the following requirements for the first and second integrals such that

$$P_i'(t_i) = P_{i+1}'(t_i) \qquad i = 1, 2, \ldots, m - 1 \tag{7.18}$$

$$P_i(t_{i-1}) = y_{i-1} \qquad P_i(t_i) = y_i \qquad i = 1, 2, \ldots, m \tag{7.19}$$

Finally we require that $P_1'(t_0) = x_2(t_0)$ and $P_m''(t_m) = 0$.

Integrating Eq. (7.17) twice, we get

$$P_i(t) = \frac{(t_i - t)^3}{6h_i}A_{i-1} + \frac{(t - t_{i-1})^3}{6h_i}A_i + \frac{a_i}{h_i}(t_i - t) + \frac{b_i}{h_i}(t - t_{i-1}) \tag{7.20}$$

where a_i and b_i are constants of integration. Substituting these values in Eqs. (7.19), we may easily solve for a_i and b_i:

$$a_i = \left(y_{j-1} - \frac{h_i^2}{6} A_{i-1} \right) \qquad b_i = \left(y_j - \frac{h_i^2}{6} A_i \right) \tag{7.21}$$

We reevaluate Eq. (7.20) to obtain

$$P_i(t) = \frac{(t_i - t)^3}{6h_i} A_{i-1} + \frac{(t - t_{i-1})^3}{6h_i} A_i + \frac{\left[y_{i-1} - (h_i^2/6) A_{i-1}\right]}{h_i} (t_i - t)$$
$$+ \frac{\left[y_i - (h_i^2/6) A_i\right]}{h_i} (t - t_{i-1}) \tag{7.22}$$

To satisfy Eq. (7.18) we take the time derivative

$$P_i'(t) = -\frac{(t_i - t)^2}{2h_i} A_{i-1} + \frac{(t - t_{i-1})^2}{2h_i} A_i$$
$$- \frac{\left[y_{i-1} - (h_i^2/6) A_{i-1}\right]}{h_i} + \frac{\left[y_i - (h_i^2/6) A_i\right]}{h_i} \tag{7.23}$$

and, therefore,

$$P_{i+1}'(t) = -\frac{(t_{i+1} - t)^2}{2h_{i+1}} A_i + \frac{(t - t_i)^2}{2h_{i+1}} A_{i+1}$$
$$- \frac{\left[y_i - (h_{i+1}^2/6) A_i\right]}{h_{i+1}} + \frac{\left[y_{i+1} - (h_{i+1}^2/6) A_{i+1}\right]}{h_{i+1}} \tag{7.24}$$

Evaluating the last two expressions at t_i, we obtain

$$P_i'(t_i) = \frac{h_i}{3} A_i + \frac{h_i}{6} A_{i-1} - \frac{y_{i-1}}{h_i} + \frac{y_i}{h_i} \tag{7.25}$$

and

$$P_{i+1}'(t_i) = -\frac{h_{i+1}}{3} A_i - \frac{h_{i+1}}{6} A_{i+1} - \frac{y_i}{h_{i+1}} + \frac{y_{i+1}}{h_{i+1}} \tag{7.26}$$

Equating Eqs. (7.25) and (7.26) we get the following $m - 1$ linear equations in terms of the $m + 1$ unknowns $A_0, A_1, \ldots, A_m$

$$\frac{h_i}{6} A_{i-1} + \left(\frac{h_i + h_{i+1}}{3} \right) A_i + \frac{h_{i+1}}{6} A_{i+1} = -\frac{y_i}{h_{i+1}} + \frac{y_{i+1}}{h_{i+1}} + \frac{y_{i-1}}{h_i} - \frac{y_i}{h_i} \tag{7.27}$$

or

$$h_i A_{i-1} + 2(h_i + h_{i+1}) A_i + h_{i+1} A_{i+1} = 6 \left(\frac{y_{i+1} - y_i}{h_{i+1}} - \frac{y_i - y_{i-1}}{h_i} \right) \tag{7.28}$$

To complete the solution we need two more equations, which are

$$P_1'(t_0) = -\frac{h_1}{3}A_0 - \frac{h_1}{6}A_1 - \frac{y_0}{h_1} + \frac{y_1}{h_1} = x_2(t_0) \tag{7.29}$$

hence

$$2h_1A_0 + h_1A_1 = 6\left(\frac{y_1 - y_0}{h_1} - x_2(t_0)\right) \tag{7.30}$$

and finally

$$P_m''(t_m) = A_m = 0 \tag{7.31}$$

Let

$$A = \begin{bmatrix} A_0 \\ A_1 \\ \cdot \\ \cdot \\ A_{m-1} \end{bmatrix} \tag{7.32}$$

$$H = \begin{bmatrix} 2h_1 & h_1 & 0 & \cdot & 0 \\ h_1 & 2(h_1+h_2) & h_2 & \cdot & \cdot \\ 0 & \cdot & \cdot & \cdot & \cdot \\ \cdot & \cdot & \cdot & \cdot & h_{m-1} \\ 0 & \cdot & \cdot & h_{m-1} & 2(h_{m-1}+h_m) \end{bmatrix} \tag{7.33}$$

and

$$W[x_1(t_0), x_2(t_0), y_1, y_2, y_3, \dots, y_m] = 6\begin{bmatrix} \dfrac{y_1 - x_1(t_0)}{h_1} - x_2(t_0) \\ \dfrac{y_2 - y_1}{h_2} - \dfrac{y_1 - x_1(t_0)}{h_1} \\ \dfrac{y_3 - y_2}{h_3} - \dfrac{y_2 - y_1}{h_2} \\ \cdot \\ \cdot \\ \dfrac{y_m - y_{m-1}}{h_m} - \dfrac{y_{m-1} - y_{m-2}}{h_{m-1}} \end{bmatrix} \tag{7.34}$$

From this definition one obtains the following relation:

$$HA = W[x_1(t_0), x_2(t_0), y_1, y_2, y_3, \dots, y_m] \tag{7.35}$$

The required control at t_0 is, therefore,

$$u(t_0) = A_0 = [1, 0, \dots, 0]H^{-1}\,W[x_1(t_0), x_2(t_0), y_1, y_2, y_3, \dots, y_m] \tag{7.36}$$

Omitting the subscript 0 (as in the previous derivations of Chapters 3 and 4) and noting that H depends on t via $h_1(t) = t_1 - t$, we obtain the feedback law

$$u(t) = [1, 0, \dots, 0]H^{-1}(t)W[x_1(t), x_2(t), y_1, y_2, y_3, \dots, y_m] \tag{7.37}$$

V. Special Cases

We now consider the special cases of a single, double, and triple target scenarios.

A. Single Target (*m* = 1)

For this case $H = [2h_1]$ and

$$W = 6\left[\frac{y_1 - x_1(t)}{h_1} - x_2(t)\right]$$

Since $y_1 = 0$, we obtain that

$$u = (1)(2h_1)^{-1}6\left[\frac{y_1 - x_1(t)}{h_1} - x_2(t)\right] = -3\left[\frac{x_1}{(t_1 - t)^2} + \frac{x_2}{(t_1 - t)}\right] \tag{7.38}$$

which is the familiar PN.

B. Double Target (*m* = 2)

For this case

$$H = \begin{bmatrix} 2h_1 & h_1 \\ h_1 & 2(h_1 + h_2) \end{bmatrix}$$

and

$$W = 6\begin{bmatrix} \dfrac{y_1 - x_1(t)}{h_1} - x_2(t) \\ \dfrac{(y_2 - y_1)}{h_2} - \dfrac{y_1 - x_1(t)}{h_1} \end{bmatrix}$$

Since $y_2 = 0$, we obtain that

$$u(t) = [1 \quad 0]\begin{bmatrix} 2h_1 & h_1 \\ h_1 & 2(h_1 + h_2) \end{bmatrix}^{-1} 6\begin{bmatrix} \dfrac{y_1 - x_1(t)}{h_1} - x_2(t) \\ \dfrac{(-y_1)}{h_2} - \dfrac{y_1 - x_1(t)}{h_1} \end{bmatrix} \tag{7.39}$$

Using a MAPLE program (Listing 7.1) we get the linear feedback form for $t < t_1$:

$$u = k_1 x_1 + k_2 x_2 + k_{y1} y_1 \tag{7.40}$$

where

$$k_1 = \frac{-6(3h_1 + 2h_2)}{h_1^2(3h_1 + 4h_2)}$$

$$k_2 = \frac{-12(h_1 + h_2)}{h_1(3h_1 + 4h_2)} \tag{7.41}$$

$$k_{y1} = \frac{6(h_1 + h_2)(h_1 + 2h_2)}{h_2 h_1^2(3h_1 + 4h_2)}$$

Listing 7.1 Feedback gains for two targets

```
> #listing 7.1
> #double target case;
> with(linalg);
> A:=array([[2*h1,h1],[h1,2*(h1+h2)]]);
> B:=array(1..2,1..1,[[(y1−x1)/h1−x2],[(−y1)/h2−(y1−x1)/h1]]);
> D:=transpose(array([[1],[0]]));
> # control evaluation;
> u:=evalm(D*inverse(A)*B);
> u1:=evalm(6*u);
```

For $t \geq t_1$ Eq. (7.38) is valid since the problem should be resolved with a single target. This result was first obtained by Norbutas.[1] Notice that for $h_1 \ll h_2$ (i.e., when the time-to-go to the second target is large enough with respect to the time-to-go to the first one) we obtain PN to y_1 by the relation

$$u = \frac{-3}{h_1^2}(x_1 - y_1) + \frac{-3}{h_1} x_2$$

C. Triple Target (m = 3)

For this case

$$H = \begin{bmatrix} 2h_1 & h_1 & 0 \\ h_1 & 2(h_1 + h_2) & h_2 \\ 0 & h_2 & 2(h_2 + h_3) \end{bmatrix}$$

and

$$W = 6 \begin{bmatrix} \dfrac{y_1 - x_1(t)}{h_1} - x_2(t) \\ \dfrac{(y_2 - y_1)}{h_2} - \dfrac{y_1 - x_1(t)}{h_1} \\ \dfrac{(y_3 - y_2)}{h_3} - \dfrac{(y_2 - y_1)}{h_3} \end{bmatrix}$$

Since $y_3 = 0$, we obtain that

$$u(t) = [1 \quad 0 \quad 0] \begin{bmatrix} 2h_1 & h_1 & 0 \\ h_1 & 2(h_1 + h_2) & h_2 \\ 0 & h_2 & 2(h_2 + h_3) \end{bmatrix}^{-1} \times 6 \begin{bmatrix} \dfrac{y_1 - x_1(t)}{h_1} - x_2(t) \\ \dfrac{(y_2 - y_1)}{h_2} - \dfrac{y_1 - x_1(t)}{h_1} \\ \dfrac{(-y_2)}{h_3} - \dfrac{(y_2 - y_1)}{h_3} \end{bmatrix} \tag{7.42}$$

Listing 7.2 Feedback gains for three targets

```
> #listing 7.2
> # triple target solution;
> with(linalg);
> A:=array([[2*h1,h1,0],[h1,2*(h1+h2),h2],[0,h2,2*(h2+h3)]]);
> B:=array([[(y1−x1)/h1−x2],[(y2−y1)/h2−(y1−x1)/h1],[(−y2)/h3−(y2−y1)/h3]]);
> D:=transpose(array([[1],[0],[0]]));
> u:=evalm(D*inverse(A)*B);
> u1:=evalm(6*");
```

Using a MAPLE program (Listing 7.2) we get the linear feedback form for $t < t_1$:

$$u = k_1 x_1 + k_2 x_2 + k_{y1} y_1 + k_{y2} y_2 \tag{7.43}$$

where

$$\begin{aligned} k_1 &= \frac{-3(6h_2h_1 + 3h_2^2 + 6h_1h_3 + 4h_2h_3)}{h_1^2(3h_1h_2 + 3h_2^2 + 3h_1h_3 + 4h_2h_3)} \\ k_2 &= \frac{-3(4h_2h_1 + 3h_2^2 + 4h_1h_3 + 4h_2h_3)}{h_1(3h_1h_2 + 3h_2^2 + 3h_1h_3 + 4h_2h_3)} \\ k_{y1} &= \frac{3(h_1 + h_2)(3h_2h_1 + 3h_2^2 + 2h_1h_3 + 4h_2h_3)}{h_2h_1^2(3h_1h_2 + 3h_2^2 + 3h_1h_3 + 4h_2h_3)} \\ k_{y2} &= \frac{-3(h_2 + h_3)(h_2 + 2h_3)}{h_2h_3(3h_1h_2 + 3h_2^2 + 3h_1h_3 + 4h_2h_3)} \end{aligned} \tag{7.44}$$

For $t_2 > t \geq t_1$ Eq. (7.40) is valid (two target case), while for $t \geq t_2$ Eq. (7.38) is the control law.

VI. Numerical Example

Consider the double target case with the following numerical values:

$$t_0 = 0 \qquad t = 0.5 \qquad t_f = 1$$

$$x_1(t_0) = 0 \qquad x_2(t_0) = 0 \qquad y_1 = 1 \qquad y_2 = 0$$

Figure 7.1 compares the MEG control with sequential PN.

The maximal control is less than 21 control units as opposed to 30 units required by PN. Figures 7.2 and 7.3 present the state time histories for both strategies. One can see how the trajectory of the MEG, while approaching the first target, takes into consideration the presence of the second one and curves the trajectory toward it. Comparing the cost of the two solutions presented as Fig. 7.4, one can see the savings by the ratio of 11:17.5 in the total cost. Notice the anticipated advantage of PN with respect to the first target only.

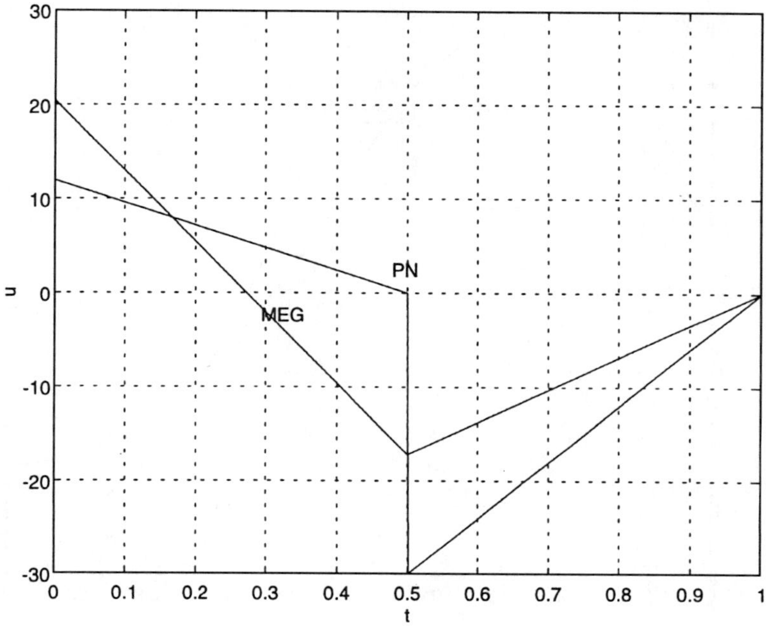

Fig. 7.1 Missile acceleration due to heading error.

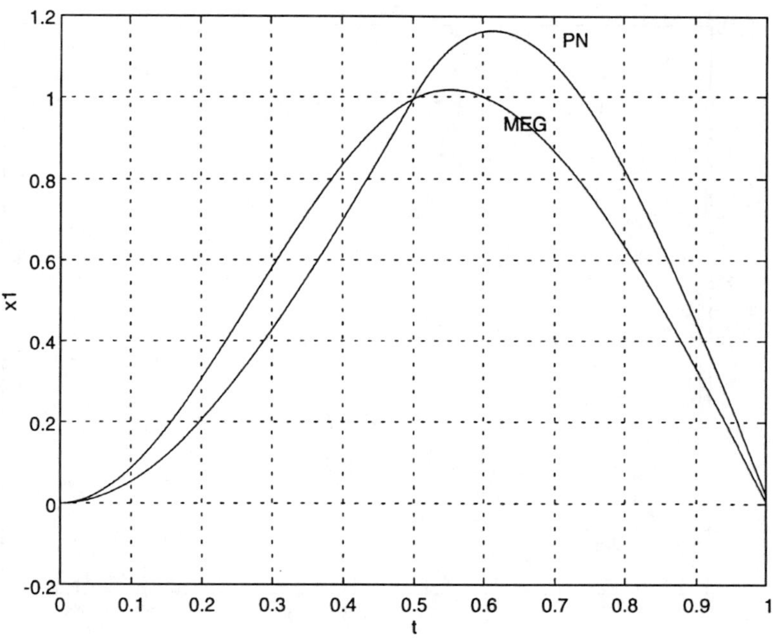

Fig. 7.2 Relative separation due to heading error.

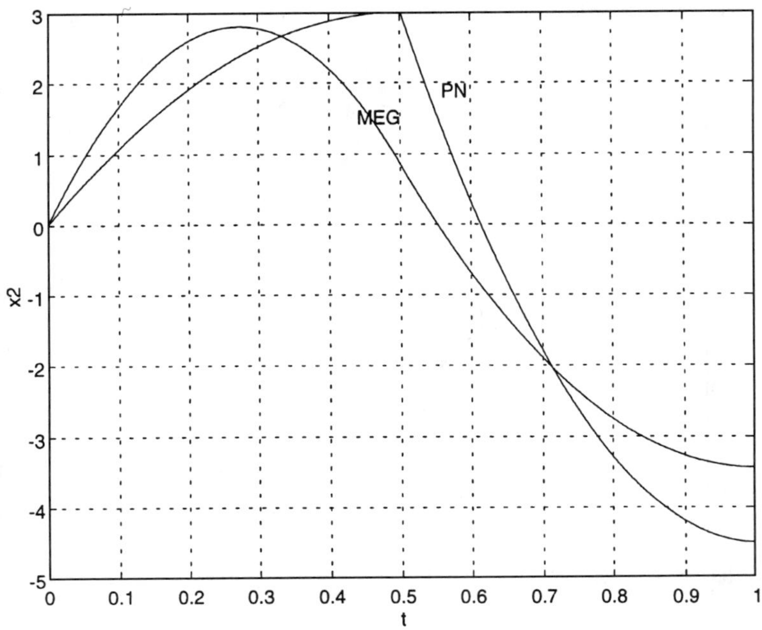

Fig. 7.3 Relative velocity due to heading error.

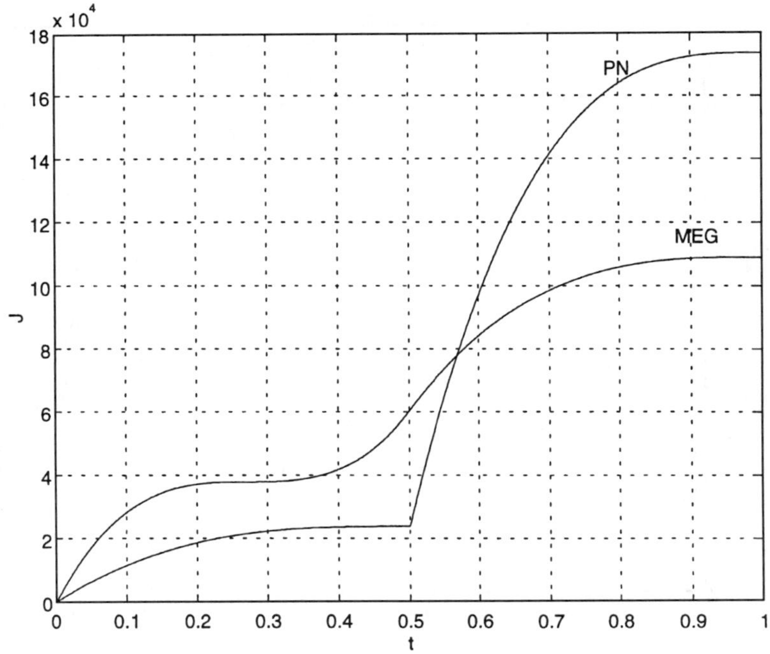

Fig. 7.4 Control effort due to heading error.

VII. Conclusions

The minimum-effort multiple-target interception problem has been solved by identifying it as a cubic-spline interpolation problem. The closed-loop control involves, in general, an on-line inversion of a $n \times n$ matrix. If, however, the number of targets is known, the inversion can be done off-line by MAPLE (or other present day symbolic codes) to obtain a simple linear feedback form.

References

[1]Norbutas, R. J., "Optimal Intercept Guidance for Multiple Target Sets," MIT Rept. ESL-R-333, Cambridge, Massachusetts, Jan. 1968, pp. 1–6.

[2]Ben-Asher, J. Z., "Minimum Effort Interception of Multiple Targets," *Journal of Guidance, Control, and Dynamics*, Vol. 16, No. 3, 1993, pp. 600–602.

[3]Bryson, A. E., and Ho, Y. C., *Applied Optimal Control*, Hemisphere, New York, 1975, Chap. 3.

[4]King, J. T., *Introduction to Numerical Computation*, McGraw-Hill, New York, 1984, Chap. 4.

Appendix A

Justification of the Sweep Assumption

THIS appendix justifies the sweep assumption of Eq. (2.20). To this end, we denote by $\phi(t, \tau)$ the transition matrix that corresponds to Eq. (2.19), namely,

$$\frac{\partial}{\partial t}\phi(t, \tau) = H\phi(t, \tau) \tag{A.1}$$

$$\phi(\tau, \tau) = I \tag{A.2}$$

where

$$H = \begin{bmatrix} A & \gamma^{-2}B_1B_1^T - B_2R^{-1}B_2^T \\ -Q & -A^T \end{bmatrix} \tag{A.3}$$

Denoting

$$B = \begin{bmatrix} B_3 \\ 0 \end{bmatrix} \quad \text{and} \quad \xi(t) = \begin{bmatrix} x^*(t) \\ \lambda(t) \end{bmatrix} \tag{A.4}$$

Eq. (2.19) becomes

$$\dot{\xi} = H(t)\xi(t) + B(t)d(t) \tag{A.5}$$

with

$$\xi(t_f) = \begin{bmatrix} x^*(t_f) \\ Q_f x^*(t_f) \end{bmatrix} \tag{A.6a}$$

$$\xi(0) = \begin{bmatrix} x^*(0) \\ \lambda(0) \end{bmatrix} = \begin{bmatrix} 0 \\ \lambda(0) \end{bmatrix} \tag{A.6b}$$

Therefore, the solution of Eq. (A.5) is of the form:

$$\xi(t) = \phi(t, t_f)\xi(t_f) + \int_{t_f}^{t} \phi(t, \tau)B(\tau)d(\tau)\, d\tau \tag{A.7}$$

where $\phi(t, \tau)$ is as previously defined.

Letting

$$\xi_1(t) = \phi(t, t_f)\xi(t_f) \equiv \begin{bmatrix} x_1^* \\ \lambda_1 \end{bmatrix} \tag{A.8}$$

and

$$\xi_2(t) = \int_{t_f}^{t} \phi(t,\tau)B(\tau)d(\tau)\,\mathrm{d}\tau \equiv \begin{bmatrix} x_2^* \\ \lambda_2 \end{bmatrix} \tag{A.9}$$

we have

$$\xi(t) = \xi_1(t) + \xi_2(t) \tag{A.10}$$

Since $\xi_2(t_f) = 0$, we obtain $\xi(t_f) = \xi_1(t_f)$ and therefore,

$$\xi_1(t) = \phi(t, t_f)\xi_1(t_f) \tag{A.11}$$

Moreover, from Eqs. (2.20), (2.25), (A.10), and the fact that $\xi_2(t_f) = 0$, we find that

$$\lambda_1(t_f) = Q_f x_1^*(t_f) \tag{A.12}$$

It follows from Eq. (A.11) that by partitioning $\phi(t, t_f)$ conformally with x_1 and λ_1 we have

$$\begin{bmatrix} x_1^*(t) \\ \lambda_1(t) \end{bmatrix} = \begin{bmatrix} \phi_{11}(t,t_f) & \phi_{12}(t,t_f) \\ \phi_{21}(t,t_f) & \phi_{22}(t,t_f) \end{bmatrix} \begin{bmatrix} x_1^*(t_f) \\ Q_f x_1^*(t_f) \end{bmatrix} \tag{A.13}$$

or equivalently

$$x_1^*(t) = [\phi_{11}(t,t_f) + \phi_{12}(t,t_f)Q_f]x_1^*(t_f) \tag{A.14}$$

$$\lambda_1(t) = [\phi_{21}(t,t_f) + \phi_{22}(t,t_f)Q_f]x_1^*(t_f) \tag{A.15}$$

If $\phi_{11}(t,t_f) + \phi_{12}(t,t_f)Q_f$ is nonsingular, then

$$\lambda_1(t) = [\phi_{21}(t,t_f) + \phi_{22}(t,t_f)Q_f][\phi_{11}(t,t_f) + \phi_{12}(t,t_f)Q_f]^{-1}x_1^*(t) \tag{A.16}$$

Denoting $P(t)$ as

$$P(t) = [\phi_{21}(t,t_f) + \phi_{22}(t,t_f)Q_f][\phi_{11}(t,t_f) + \phi_{12}(t,t_f)Q_f]^{-1} \tag{A.17}$$

we get

$$\lambda_1(t) = P(t)x_1^*(t) \tag{A.18}$$

It can readily be verified that $P(t)$ of Eq. (A.17) is the solution to Eqs. (2.29a) and (2.29b).

From Eq. (A.10) we then have

$$\begin{aligned} \lambda(t) &= \lambda_1(t) + \lambda_2(t) = P(t)x_1^*(t) + \lambda_2(t) \\ &= P(t)[x_1^*(t) + x_2^*(t)] + \lambda_2(t) - P(t)x_2^*(t) \end{aligned} \tag{A.19}$$

However, from Eqs. (A.9) and (A.10) we find the following expressions:

$$x_2^*(t) = \int_{t_f}^{t} \phi_{11}(t,\tau)B_3(\tau)d(\tau)\,\mathrm{d}\tau \tag{A.20}$$

and

$$\lambda_2(t) = \int_{t_f}^{t} \phi_{21}(t,\tau)B_3(\tau)d(\tau)\,\mathrm{d}\tau \tag{A.21}$$

Therefore, we have from Eqs. (A.19) and (A.20–A.21) that

$$\lambda(t) = P(t)x^*(t) + \theta_1(t) \tag{A.22}$$

where

$$\theta_1(t) = \int_{t_f}^{t} [\phi_{21}(t, \tau) - P(t)\phi_{11}(t, \tau)]B_3(\tau)d(\tau)\, \mathrm{d}\tau \tag{A.23}$$

which proves our sweep assumption.

Appendix B

Mathematical Computations with MAPLE

I. General

MAPLE™ V is a powerful tool for carrying out symbolic mathematical computations. This package is extensively used in our monograph. To facilitate its use we present a brief description of MAPLE V, emphasizing the main commands that are frequently used in this book. For more details the interested reader is referred to Ref. 1.

II. Program Description and Its Use in This Book

MAPLE V is intended for mathematical computations—numerical, graphical, and most importantly, symbolic. The latter feature is the one used in the present book. MAPLE V runs on many types of computer systems, but all versions are essentially the same computationally and they differ mainly in their user interfaces. The version of the accompanying software is compatible with MS Windows 3.xx. The programs are run interactively by entering the commands line by line.

We use this program in the present book mainly to solve algebraic and differential equations analytically. Consequently, we shall concentrate in the following on explaining the features that help to solve these algebraic and differential equations.

III. Assigning and Unassigning Variables

First we need to know how to directly assign a value to a variable. This is done simply by the command:

variable:=expression;

where the variable is a symbol and the expression may be numeric or symbolic (using other symbols—whether previously assigned or unassigned). For example *a:=b*c^2*; (the variable *a* from here on) becomes bc^2.

Remark: MAPLE V has preassigned numerical values for the following variables: pi $= \pi$, $I = \sqrt{-1}$, $E = e$ (natural log base), infinity $= \infty$, and gamma $= \gamma$ (Euler's constant). For this reason we use β in our differential games examples, although the text uses γ as is common for most textbooks on differential games.

Another way of assigning a value to a variable is indirectly by solving an equation for which the variable is the unknown. The actual assignment is done by explicitly commanding *assign(");* after solving the equation. For example,

solve(3bx+4=0,x);

assign(");.

The variable x will be assigned the value $-4/(3b)$. To unassign the variable x one can use the command

x:='x';,

by which x again becomes mathematically unknown.

Finally, it is always advisable to try to simplify an expression by using

simplify(expression);

For example, *simplify(a*a/a);* (yielding a).

IV. Evaluating Expressions

As already explained, an assignment command changes the value of a variable, such that this variable—unless unassigned—will automatically be replaced by its assignment. We may, however, evaluate variables or expressions of several variables without necessarily changing their values.

To this end we shall briefly describe three frequently used commands.

1) To substitute for a variable in an expression we use

subs(variable=replacement, expression);

Example: *subs(x=4, x^2+b)*; (yielding *16+b*).

2) To find a limit value of an expression we use:

limit(function, variable=limit value);

Example: *limit (1/t,t=infinity)*; (yielding *0*).

3) To evaluate numerically an expression using floating-point numbers, the following command is used:

evalf(expression);

Example: *evalf(Pi);* (yielding 3.141592654).

V. Solving Equations

The main feature of MAPLE V exploited in this book is its surprisingly powerful capability of solving both algebraic and differential equations. The following is the simplest way of perfoming these tasks.

1) To solve an algebraic equation the following command is used:

solve(equation, variable);

Example: *solve(x^2−2*x+b,x);* (yielding $1+\sqrt{1-b}$ and $1-\sqrt{1-b}$).

Similarly, solving a set of algebraic equations is performed by

solve(set of equations, set of variables);

Example: *solve({x+y+3,4*x+y−6}, {x, y})*; (yielding *{y=−6, x=3}*).

2) To solve a differential equation we first need to present the differentiation operator:

diff(expression,variable for differentiation);

Example: *diff*(x^2, x); (yielding $2x$).
For higher order differentiations use *$n* as follows:

diff(expression,variable$n);

Example: *diff(x^2,x$2)*; (yielding 2).

Simply to integrate a function one can use *int(f(t));*, which yields the indefinite integral of f. Using *int(f(t),t=0..T);* yields the definite integral with the specified boundaries.
Examples:

int(sin(x), x); [yielding - cos(x)]

int(sin(x), x=0..Pi); (yielding 2).

Differential equations can be solved by

dsolve(set of equations and initial conditions, set of variables);

Example:

dsolve({diff(x1(t),t)=x2(t),diff(x2(t),t)=1,x1(0)=0,x2(0)=0}, {x1(t),x2(t)});

(yielding $\{x1(t) = \frac{1}{2}t^2, x2(t) = t\}$).

Recall that the solution values are not assigned automatically. We may assign them (if we wish to do so) by commanding *assign(")*; right after solving the equations.
Example:

dsolve({diff(y(t),t)−y(t)=0,y(0)=1},y(t));

assign(");

The variable y will be assigned the value e^t.

VI. Linear Algebra

Other features of MAPLE used in this monograph (mainly to handle matrix Riccati equations) are related to linear algebra. To call the functions one must first load the linear algebra package by typing the following command:

with(linalg);

The commonly used commands are 1) defining an array:

variable:=array([ ]);

Example: *A:=array([[1,0],[0,1]]);* yielding

$$A = \begin{bmatrix} 1 & 0 \\ 0 & 1 \end{bmatrix}$$

2) evaluating a matrix using matrix operations:

variable:=evalm(expression);

Example: *A:=evalm(B+C);* (yielding A=B+C).
3) transposing a matrix:

variable:=transpose(variable);

Example: *A:=transpose(B);* (yielding $A = B^T$).
4) matrix multiplication:

variable:=evalm(variable & variable)*;

Example: *A:=evalm(B&*C);* (yielding A=BC).
5) matrix inversion:

variable:=inverse(variable);

Example: *A:=inverse(B); (yielding* $A = B^{-1}$).
and 6) eigenvalues:

variable:=eigenval(variable);

Example: *r:=eigenval(array([[1,0],[0,1]]));* (yielding {1,1}).

VII. Representative Examples

A. Example B1: Unconstrained Function Minimizing

We want to minimize the following function

$$f(x) = e^x(4x^2 + 2y^2 + 4xy + 2y + 1)$$

This example was employed by Ref. 2 to demonstrate a numerical optimization routine. In Listing B.1 we solve it analytically.

B. Example B2: Adjoint Analysis for the Example of Chapter 2

In the adjoint analysis example of Chapter 2 (Example 2.3, case 1) we had to solve for the following relations:

$$\frac{\mathrm{d}\bar{\lambda}(\tau)}{\mathrm{d}\tau} = -\frac{\bar{\lambda}(\tau)}{(1/c) + \tau} \qquad \bar{\lambda}(0) = 1$$

$$x(t_f) = \int_0^{t_f} \bar{\lambda}(t)\,\mathrm{d}t$$

This task is easily done by the following commands:

```
>sys:=(diff(y(t),t)+y(t)/(1/c+(1)*t));
>dsolve({sys,y(0)=1},y(t));
>assign('');
>int(y(t),t=0..tf)
```

Listing B.1 Function minimization

```
# We first define the cost function f(x,y)
> f(x,y):=exp(x)*(4*x^2+2*y^2+4*x*y+2*y+1);
# Then we evaluate the derivatives:
> dx:=diff(f(x,y),x);
> dy:=diff(f(x,y),y);
# To find stationary points we solve:
> solve({dx,dy},{x,y});
# We obtain two solutions {−3/2,1} and {1/2,−1};
# We evaluate the Hessain
>h11:=diff(dx,x);
>h12:=diff(dx,y);
>h21:=diff(dy,x);
>h22:=diff(dy,y);
>with(linalg);
>H:=array([[h11,h12],[h21,h22]]);
# And we calculate the eigenvalues:
>r:=eigenvals(H);
# We evaluate the result at {−3/2,1}
>r1:=evalf(subs(x=−3/2,y=1,[r]));
# Resulting eigenvalues are {1.44,−.55} - hence {−3/2,1} is a saddle point.
>We evaluate the e.v.'s at {1/2,−1}
> r2:=evalf(subs(x=1/2,y=−1,[r]));
# We obtain {17.3,2.5} - hence {1/2,−1} is a minimum point.
# The minimum cost is zero, as can be obtained from
>evalf(subs(x=1/2,y=−1,[f(x,y)]);.
```

The result is

$$\frac{1}{c}\ln(1+ct_f)$$

VIII. Concluding Remarks

A very brief introduction to MAPLE V was presented covering only the topics that are used in this book. For more details the interested reader is referred to Ref. 1. The MAPLE files in the accompanying diskette (subdirectory MAP) are exactly as listed in the book with the listing numbers serving as the file names and with the extension of .ms (e.g., listing31.ms is the MAPLE file of the program given in Listing 3.1).

References

[1]Char, B. W., Geddes, K. O., Gonnet, G. H., Leong, B. L., Monagan, M. B., and Walt, S. M., *"First Leaves: A Tutorial Introduction to MAPLE V,* Springer-Verlag, New York, 1992.

[2]Grace, A., *Optimization Toolbox User's Guide,* The MathWorks, Inc., Natick, MA, 1990, Chap. 2.

Numerical Computations with MATLAB

I. Scope

MATLAB™ is a technical computing environment for high performance numerical computation and visualization. The fact that MATLAB integrates matrix computation and graphics (beside many other features that we do not use) makes it a natural tool for the numerical simulations that were used for the examples in this book.

This appendix provides a brief introduction to the MATLAB environment, describing the most important computational features as well as the most useful input/output commands. As a demonstrative example, it also provides a description of the simulation model used to produce the numerical results for missiles modeled by higher order dynamics of Chapter 6. This model can be used to produce further simulation results by minor adjustments.

II. Brief Introduction to MATLAB

The details in this section provide a brief introduction to MATLAB; a full introduction is available in Ref. 1. MATLAB uses the assignment "=", which implies assignment of an expression to a variable.

```
x = 2 * a;
```

means assignment of $2a$ into x. The " ; " at the end of line suppresses the display of the assignment result. MATLAB naturally deals with matrices, where vectors and scalars can be viewed as special cases of matrices. Matrices are defined within brackets "[ ]"; for example,

```
A = [ 1 2 3;4 5 6];
```

and

```
A = [ 1 2 3
      4 5 6];
```

both define the same matrix

$$A = \begin{bmatrix} 1 & 2 & 3 \\ 4 & 5 & 6 \end{bmatrix}$$

where semicolons or carriage returns separate rows. No dimension or type statements are needed. Matrices can be redefined within one program or working session. For example, the sequence of statements

A = [1 2;3 4];

A = [1 2 3;4 5 6];

is legal and the value of *A* is changed accordingly.

The basic operations are + , - ,* , /, and ^ denoting addition, subtraction, multiplication, division, and raising to a power, respectively.

*x = ((2*a) / b + c)^d;*

denotes

$$x = (2a/b + c)^d$$

where again the " ; " suppresses the display. The common mathematical functions have obvious names. For example,

y = sin(x);

y = cos(x);

y = tan(x);

y = asin(x);

y = acos(x);

y = atan(x);

apply the functions sine, cosine, tangent, arcsine, arccosine, and arctangent, respectively, to the elements of a matrix x where angles are expressed in radians. The function

y = sqrt(x);

calculates the square root of the elements of x, whereas

y = sqrtm(x);

produces *a y*, so that $y^T y = x$. Operations between matrices are also naturally dealt with, where, for example,

A =[1 2;3 4];

B = [5 6;7 8];

*C = A * B;*

denotes the matrix multiplication of B by A.

The matrix transpose is formed with apostrophe " ' ".

B = A';

assigns the transpose of A into B; MATLAB allows also array operations on an element by element basis; for example,

C = A.+B;

C = A.−B'

*C = A.*B;*

C = A./B;
C=A.^B;

form an element-by-element addition, subtraction, multiplication, division, and power operations, respectively, between matrices A and B where the result is assigned to a matrix C. Element-by-element or regular matrix operations with inconsistent dimension of the operands result in error messages from MATLAB.

MATLAB also provides very advanced tools for graphical visualization. In our application we use the very basic tools, such as

plot(t,y);

to plot an array y versus an array t. We label our plot using, for example,

xlabel('Time [sec]');
ylabel('Position [meter]');
title('Relative Position - Proportional Navigation');

and add grid by just typing

grid;

When we want to add labels based on some meaningful numerical data to our graphs, we use either statements of the type:

title(['Relative Position - Proportional Navigation , NTAG = ',...
num2str(ntag)]);

where *ntag* is a variable in the MATLAB workspace, and *num2str(ntag)* converts the number *ntag* to a string, or equivalently,

tit1 = sprintf('Relative Position - Proportional Navigation ,...
NTAG % g',ntag); title (tit 1);

In both cases "..." signifies a line continuation. The reader is referred to the MATLAB reference guide for more advanced uses of "*sprintf*".

The graphic window can be subdivided by the command "*subplot*". For example to divide the graphic window into a 2 × 2 graph of *x1* versus *y1*, *x2* versus *y2*, etc., we use

subplot(221);
plot(x1,y1);
subplot(222);
plot(x2,y2);
subplot(223);
plot(x3,y3);
subplot(224);
plot(x4,y4);

Also, different line styles may be used.

plot(x1,y1,'–',x2,y2,':');

plots *x1* versus *y1* with a solid line, and *x2* versus *y2* with a dotted line.

All MATLAB operations can be performed either interactively at the MATLAB command prompt or by using script files that are called "m-files". The latter are just ASCII files created by text editors. An m-file should have extension ".m", (e.g., guidemaster.m).

III. Demonstrative Example

In this section we describe how simulations of cases that involve higher order dynamics for the pursuer and evader are performed. To this end, we first describe the state augmentation procedure we use to include such dynamics in the pursuit-evasion conflict model.

The pursuer's dynamic model is described by

$$\frac{dx_p}{dt} = A_p x_p + B_p u$$

$$y_p = C_p x_p + D_p u$$

where x_p is the state vector of the pursuer's dynamics, u is its acceleration command, and y_p is the pursuer's acceleration. Similarly, the evader is characterized by

$$\frac{dx_e}{dt} = A_e x_e + B_e w$$

$$y_e = C_e x_e + D_e w$$

where x_e is the state vector of the evader's dynamics, w is its acceleration command, and y_e is the evader's acceleration. We define the augmented state vector of the system by

$$\zeta = \begin{bmatrix} x_1 \\ x_2 \\ x_p \\ x_e \end{bmatrix}$$

where the evolution of the relative position is given by

$$\frac{dx_1}{dt} = x_2$$

and the evolution of the relative velocity is described by

$$\frac{dx_2}{dt} = y_p + y_e = C_p x_p + C_e x_e + D_p u + D_e w$$

Collecting these equations together results in our augmented system that forms the basis for simulation programs for higher order dynamics. These programs typically include the building of the pursuer's and evader's transfer functions by using the MATLAB's "*conv*" and "*tf2ss*" commands (note that the MATLAB Control Toolbox is required for this purpose). For example, to define the pursuer and evader models

$$G_p(s) = (\tau s + 1)\omega_n^2/(s^2 + 2\zeta w_n s + w_n^2), \qquad G_e(s) = 1/(0.05\,S + 1)$$

with $\tau = 0.5$, $\zeta = 0.707$, $w_n = 10$, we may use the sequence of commands of Listing C.1. Three simulation programs that employ the model of Listing C.1 are given in the accompanying diskette. This programs are sim_appb_1.m : forward simulation; sim_appb_2.m : miss-distance adjoint simulation; sim_appb_3.m : hit-angle adjoint simulation. These programs were used to produce the results of Chapter 6 (Sec. V.C), which deals with higher order dynamics, but are also most suitable to work additional examples by making a few minor adjustments that involve commenting and uncommenting of command lines in these programs (by adding or removing "%" at the line beginning).

Listing C.1 Higher order model

```
tau  = 0.5;
wn  = 10;
ksi  = 0.707;
Nump =  [0 1];
Denp  = [tau 1];
Nump2 = [0 0 wn^2];
Denp2 = [1 2*ksi*wn wn^2];
Nump  = conv(Nump,Nump2);
Denp  = conv(Denp,Denp2);
[Ap,Bp,Cp,Dp] = tf2ss(Nump,Denp);   % form   state-space description from
                                              transfer function
Nume = [0 1];
Dene = [0.05 1];
[Ae,Be,Ce,De] = t+255 (Nume, Dene)
rp = length(Ap);
re = length(Ae);
A = [0            1             zeros(1,rp)   zeros(1,re)
     0            0             Cp            Ce
   zeros(rp,1)  zeros(rp,1)   Ap            zeros(rp,re)
   zeros(re,1)  zeros(re,1)   zeros(re,rp)  Ae];
B = [ 0
     Dp
     Bp
     zeros(re,1)];
D = [0
     De
     zeros(rp,1)
     Be];
```

IV. Usage of the MATLAB Programs

The MATLAB programs accompanying this book reside in the subdirectory MAT in the attached diskette. The programs that serve as main programs (are called from the MATLAB prompt) have the names of the *corresponding tables* in the text with the extension of .m. However, programs that are called from these main programs have meaningful names, rather than the *table names*. For example, the program of Listing 3.12 is named "prop_nav_gains". For additional details and tips and program usage examples, see the file "readme.m" in the subdirectory MAT.

V. Concluding Remarks

This appendix has provided some basic and useful information about MATLAB and more information can be obtained by contacting The Mathworks, Inc., 24 Prime Park Way, Natick, MA 01760 (phone: (508) 653-1415; Fax: (508) 653-2997; E-mail: info@mathworks.com).

The programs in the accompanying diskette may be used to reproduce the numerical results of this book as well as to produce new results and work out new examples. For example, the transfer function *nump(s)/denp(s)* of the pursuer may be obtained from the design of an acceleration loop with one of the MATLAB control related toolboxes for some open-loop models of a pursuer. Thus, these programs can be used for generating various projects for students.

Reference

[1]MATLAB *High Performance Numeric Computation and Visualization Software—Reference Guide*, The MathWorks, Inc., Natick, MA, 1995.

Index

PROGRESS IN ASTRONAUTICS AND AERONAUTICS SERIES VOLUMES

***1. Solid Propellant Rocket Research (1960)**
Martin Summerfield
Princeton University

***2. Liquid Rockets and Propellants (1960)**
Loren E. Bollinger
Ohio State University
Martin Goldsmith
The Rand Corp.
Alexis W. Lemmon Jr.
Battelle Memorial Institute

***3. Energy Conversion for Space Power (1961)**
Nathan W. Snyder
Institute for Defense Analyses

***4. Space Power Systems (1961)**
Nathan W. Snyder
Institute for Defense Analyses

***5. Electrostatic Propulsion (1961)**
David B. Langmuir
Space Technology Laboratories, Inc.
Ernst Stuhlinger
NASA George C. Marshall Space Flight Center
J. M. Sellen Jr.
Space Technology Laboratories, Inc.

***6. Detonation and Two-Phase Flow (1962)**
S. S. Penner
California Institute of Technology
F. A. Williams
Harvard University

***7. Hypersonic Flow Research (1962)**
Frederick R. Riddell
AVCO Corp.

***8. Guidance and Control (1962)**
Robert E. Roberson
Consultant
James S. Farrior
Lockheed Missiles and Space Co.

***9. Electric Propulsion Development (1963)**
Ernst Stuhlinger
NASA George C. Marshall Space Flight Center

***10. Technology of Lunar Exploration (1963)**
Clifford I. Cumming
Harold R. Lawrence
Jet Propulsion Laboratory

***11. Power Systems for Space Flight (1963)**
Morris A. Zipkin
Russell N. Edwards
General Electric Co.

***12. Ionization in High-Temperature Gases (1963)**
Kurt E. Shuler, Editor
National Bureau of Standards
John B. Fenn, Associate Editor
Princeton University

***13. Guidance and Control–II (1964)**
Robert C. Langford
General Precision Inc.
Charles J. Mundo
Institute of Naval Studies

***14. Celestial Mechanics and Astrodynamics (1964)**
Victor G. Szebehely
Yale University Observatory

***15. Heterogeneous Combustion (1964)**
Hans G. Wolfhard
Institute for Defense Analyses
Irvin Glassman
Princeton University
Leon Green Jr.
Air Force Systems Command

***16. Space Power Systems Engineering (1966)**
George C. Szego
Institute for Defense Analyses
J. Edward Taylor
TRW Inc.

***17. Methods in Astrodynamics and Celestial Mechanics (1966)**
Raynor L. Duncombe
U.S. Naval Observatory
Victor G. Szebehely
Yale University Observatory

***18. Thermophysics and Temperature Control of Spacecraft and Entry Vehicles (1966)**
Gerhard B. Heller
NASA George C. Marshall Space Flight Center

***19. Communication Satellite Systems Technology (1966)**
Richard B. Marsten
Radio Corporation of America

*Out of print.

***20. Thermophysics of Spacecraft and Planetary Bodies: Radiation Properties of Solids and the Electromagnetic Radiation Environment in Space (1967)**
Gerhard B. Heller
NASA George C. Marshall Space Flight Center

***21. Thermal Design Principles of Spacecraft and Entry Bodies (1969)**
Jerry T. Bevans
TRW Systems

***22. Stratospheric Circulation (1969)**
Willis L. Webb
Atmospheric Sciences Laboratory, White Sands, and University of Texas at El Paso

***23. Thermophysics: Applications to Thermal Design of Spacecraft (1970)**
Jerry T. Bevans
TRW Systems

***24. Heat Transfer and Spacecraft Thermal Control (1971)**
John W. Lucas
Jet Propulsion Laboratory

25. Communication Satellites for the 70's: Technology (1971)
Nathaniel E. Feldman
The Rand Corp.
Charles M. Kelly
The Aerospace Corp.

26. Communication Satellites for the 70's: Systems (1971)
Nathaniel E. Feldman
The Rand Corp.
Charles M. Kelly
The Aerospace Corp.

27. Thermospheric Circulation (1972)
Willis L. Webb
Atmospheric Sciences Laboratory, White Sands, and University of Texas at El Paso

28. Thermal Characteristics of the Moon (1972)
John W. Lucas
Jet Propulsion Laboratory

***29. Fundamentals of Spacecraft Thermal Design (1972)**
John W. Lucas
Jet Propulsion Laboratory

***30. Solar Activity Observations and Predictions (1972)**
Patrick S. McIntosh
Murray Dryer
Environmental Research Laboratories, National Oceanic and Atmospheric Administration

31. Thermal Control and Radiation (1973)
Chang-Lin Tien
University of California at Berkeley

***32. Communications Satellite Systems (1974)**
P. L. Bargellini
COMSAT Laboratories

33. Communications Satellite Technology (1974)
P. L. Bargellini
COMSAT Laboratories

***34. Instrumentation for Airbreathing Propulsion (1974)**
Allen E. Fuhs
Naval Postgraduate School
Marshall Kingery
Arnold Engineering Development Center

35. Thermophysics and Spacecraft Thermal Control (1974)
Robert G. Hering
University of Iowa

36. Thermal Pollution Analysis (1975)
Joseph A. Schetz
Virginia Polytechnic Institute
ISBN 0-915928-00-0

***37. Aeroacoustics: Jet and Combustion Noise; Duct Acoustics (1975)**
Henry T. Nagamatsu, Editor
General Electric Research and Development Center
Jack V. O'Keefe, Associate Editor
The Boeing Co.
Ira R. Schwartz, Associate Editor
NASA Ames Research Center
ISBN 0-915928-01-9

***38. Aeroacoustics: Fan, STOL, and Boundary Layer Noise; Sonic Boom; Aeroacoustics Instrumentation (1975)**
Henry T. Nagamatsu, Editor
General Electric Research and Development Center
Jack V. O'Keefe, Associate Editor
The Boeing Co.
Ira R. Schwartz, Associate Editor
NASA Ames Research Center
ISBN 0-915928-02-7

39. Heat Transfer with Thermal Control Applications (1975)
M. Michael Yovanovich
University of Waterloo
ISBN 0-915928-03-5

*Out of print.

***40. Aerodynamics of Base Combustion (1976)**
S. N. B. Murthy, Editor
J. R. Osborn,
Associate Editor
Purdue University
A. W. Barrows
J. R. Ward,
Associate Editors
Ballistics Research Laboratories
ISBN 0-915928-04-3

41. Communications Satellite Developments: Systems (1976)
Gilbert E. LaVean
Defense Communications Agency
William G. Schmidt
CML Satellite Corp.
ISBN 0-915928-05-1

***42. Communications Satellite Developments: Technology (1976)**
William G. Schmidt
CML Satellite Corp.
Gilbert E. LaVean
Defense Communications Agency
ISBN 0-915928-06-X

***43. Aeroacoustics: Jet Noise, Combustion and Core Engine Noise (1976)**
Ira R. Schwartz, Editor
NASA Ames Research Center
Henry T. Nagamatsu,
Associate Editor
General Electric Research and Development Center
Warren C. Strahle,
Associate Editor
Georgia Institute of Technology
ISBN 0-915928-07-8

***44. Aeroacoustics: Fan Noise and Control; Duct Acoustics; Rotor Noise (1976)**
Ira R. Schwartz, Editor
NASA Ames Research Center
Henry T. Nagamatsu,
Associate Editor
General Electric Research and Development Center
Warren C. Strahle,
Associate Editor
Georgia Institute of Technology
ISBN 0-915928-08-6

***45. Aeroacoustics: STOL Noise; Airframe and Airfoil Noise (1976)**
Ira R. Schwartz, Editor
NASA Ames Research Center
Henry T. Nagamatsu,
Associate Editor
General Electric Research and Development Center
Warren C. Strahle,
Associate Editor
Georgia Institute of Technology
ISBN 0-915928-09-4

***46. Aeroacoustics: Acoustic Wave Propagation; Aircraft Noise Prediction; Aeroacoustic Instrumentation (1976)**
Ira R. Schwartz, Editor
NASA Ames Research Center
Henry T. Nagamatsu,
Associate Editor
General Electric Research and Development Center
Warren C. Strahle,
Associate Editor
Georgia Institute of Technology
ISBN 0-915928-10-8

***47. Spacecraft Charging by Magnetospheric Plasmas (1976)**
Alan Rosen
TRW Inc.
ISBN 0-915928-11-6

48. Scientific Investigations on the Skylab Satellite (1976)
Marion I. Kent
Ernst Stuhlinger
NASA George C. Marshall Space Flight Center
Shi-Tsan Wu
University of Alabama
ISBN 0-915928-12-4

49. Radiative Transfer and Thermal Control (1976)
Allie M. Smith
ARO Inc.
ISBN 0-915928-13-2

***50. Exploration of the Outer Solar System (1976)**
Eugene W. Greenstadt
TRW Inc.
Murray Dryer
National Oceanic and Atmospheric Administration
Devrie S. Intriligator
University of Southern California
ISBN 0-915928-14-0

***51. Rarefied Gas Dynamics, Parts I and II (two volumes) (1977)**
J. Leith Potter
ARO Inc.
ISBN 0-915928-15-9

***52. Materials Sciences in Space with Application to Space Processing (1977)**
Leo Steg
General Electric Co.
ISBN 0-915928-16-7

*Out of print.

***53. Experimental Diagnostics in Gas Phase Combustion Systems (1977)**
Ben T. Zinn, Editor
Georgia Institute of Technology
Craig T. Bowman, Associate Editor
Stanford University
Daniel L. Hartley, Associate Editor
Sandia Laboratories
Edward W. Price, Associate Editor
Georgia Institute of Technology
James G. Skifstad, Associate Editor
Purdue University
ISBN 0-915928-18-3

54. Satellite Communication: Future Systems (1977)
David Jarett
TRW Inc.
ISBN 0-915928-18-3

***55. Satellite Communications: Advanced Technologies (1977)**
David Jarett
TRW Inc.
ISBN 0-915928-19-1

***56. Thermophysics of Spacecraft and Outer Planet Entry Probes (1977)**
Allie M. Smith
ARO Inc.
ISBN 0-915928-20-5

***57. Space-Based Manufacturing from Nonterrestrial Materials (1977)**
Gerald K. O'Neill, Editor
Brian O'Leary, Assistant Editor
Princeton University
ISBN 0-915928-21-3

***58. Turbulent Combustion (1978)**
Lawrence A. Kennedy
State University of New York at Buffalo
ISBN 0-915928-22-1

***59. Aerodynamic Heating and Thermal Protection Systems (1978)**
Leroy S. Fletcher
University of Virginia
ISBN 0-915928-23-X

***60. Heat Transfer and Thermal Control Systems (1978)**
Leroy S. Fletcher
University of Virginia
ISBN 0-915928-24-8

***61. Radiation Energy Conversion in Space (1978)**
Kenneth W. Billman
NASA Ames Research Center
ISBN 0-915928-26-4

***62. Alternative Hydrocarbon Fuels: Combustion and Chemical Kinetics (1978)**
Craig T. Bowman
Stanford University
Jorgen Birkeland
Department of Energy
ISBN 0-915928-25-6

***63. Experimental Diagnostics in Combustion of Solids (1978)**
Thomas L. Boggs
Naval Weapons Center
Ben T. Zinn
Georgia Institute of Technology
ISBN 0-915928-28-0

***64. Outer Planet Entry Heating and Thermal Protection (1979)**
Raymond Viskanta
Purdue University
ISBN 0-915928-29-9

***65. Thermophysics and Thermal Control (1979)**
Raymond Viskanta
Purdue University
ISBN 0-915928-30-2

***66. Interior Ballistics of Guns (1979)**
Herman Krier
University of Illinois at Urbana–Champaign
Martin Summerfield
New York University
ISBN 0-915928-32-9

***67. Remote Sensing of Earth from Space: Role of "Smart Sensors" (1979)**
Roger A. Breckenridge
NASA Langley Research Center
ISBN 0-915928-33-7

68. Injection and Mixing in Turbulent Flow (1980)
Joseph A. Schetz
Virginia Polytechnic Institute and State University
ISBN 0-915928-35-3

---

*Out of print.

***69. Entry Heating and Thermal Protection (1980)**
Walter B. Olstad
NASA Headquarters
ISBN 0-915928-38-8

***70. Heat Transfer, Thermal Control, and Heat Pipes (1980)**
Walter B. Olstad
NASA Headquarters
ISBN 0-915928-39-6

***71. Space Systems and Their Interactions with Earth's Space Environment (1980)**
Henry B. Garrett
Charles P. Pike
Hanscom Air Force Base
ISBN 0-915928-41-8

***72. Viscous Flow Drag Reduction (1980)**
Gary R. Hough
Vought Advanced Technology Center
ISBN 0-915928-44-2

***73. Combustion Experiments in a Zero-Gravity Laboratory (1981)**
Thomas H. Cochran
NASA Lewis Research Center
ISBN 0-915928-48-5

***74. Rarefied Gas Dynamics, Parts I and II (two volumes) (1981)**
Sam S. Fisher
University of Virginia
ISBN 0-915928-51-5

***75. Gasdynamics of Detonations and Explosions (1981)**
J. R. Bowen
University of Wisconsin at Madison
N. Manson
Universite de Poitiers
A. K. Oppenheim
University of California at Berkeley
R. I. Soloukhin
Institute of Heat and Mass Transfer, BSSR Academy of Sciences
ISBN 0-915928-46-9

76. Combustion in Reactive Systems (1981)
J. R. Bowen
University of Wisconsin at Madison
N. Manson
Universite de Poitiers
A. K. Oppenheim
University of California at Berkeley
R. I. Soloukhin
Institute of Heat and Mass Transfer, BSSR Academy of Sciences
ISBN 0-915928-47-7

***77. Aerothermodynamics and Planetary Entry (1981)**
A. L. Crosbie
University of Missouri-Rolla
ISBN 0-915928-52-3

***78. Heat Transfer and Thermal Control (1981)**
A. L. Crosbie
University of Missouri-Rolla
ISBN 0-915928-53-1

***79. Electric Propulsion and Its Applications to Space Missions (1981)**
Robert C. Finke
NASA Lewis Research Center
ISBN 0-915928-55-8

***80. Aero-Optical Phenomena (1982)**
Keith G. Gilbert
Leonard J. Otten
Air Force Weapons Laboratory
ISBN 0-915928-60-4

***81. Transonic Aerodynamics (1982)**
David Nixon
Nielsen Engineering & Research, Inc.
ISBN 0-915928-65-5

***82. Thermophysics of Atmospheric Entry (1982)**
T. E. Horton
University of Mississippi
ISBN 0-915928-66-3

***83. Spacecraft Radiative Transfer and Temperature Control (1982)**
T. E. Horton
University of Mississippi
ISBN 0-915928-67-1

84. Liquid-Metal Flows and Magneto-hydrodynamics (1983)
H. Branover
Ben-Gurion University of the Negev
P. S. Lykoudis
Purdue University
A. Yakhot
Ben-Gurion University of the Negev
ISBN 0-915928-70-1

*Out of print.

***85. Entry Vehicle Heating and Thermal Protection Systems: Space Shuttle, Solar Starprobe, Jupiter Galileo Probe (1983)**
Paul E. Bauer
McDonnell Douglas Astronautics Co.
Howard E. Collicott
The Boeing Co.
ISBN 0-915928-74-4

***86. Spacecraft Thermal Control, Design, and Operation (1983)**
Howard E. Collicott
The Boeing Co.
Paul E. Bauer
McDonnell Douglas Astronautics Co.
ISBN 0-915928-75-2

***87. Shock Waves, Explosions, and Detonations (1983)**
J. R. Bowen
University of Washington
N. Manson
Universite de Poitiers
A. K. Oppenheim
University of California at Berkeley
R. I. Soloukhin
Institute of Heat and Mass Transfer, BSSR Academy of Sciences
ISBN 0-915928-76-0

88. Flames, Lasers, and Reactive Systems (1983)
J. R. Bowen
University of Washington
N. Manson
Universite de Poitiers
A. K. Oppenheim
University of California at Berkeley
R. I. Soloukhin
Institute of Heat and Mass Transfer, BSSR Academy of Sciences
ISBN 0-915928-77-9

***89. Orbit-Raising and Maneuvering Propulsion: Research Status and Needs (1984)**
Leonard H. Caveny
Air Force Office of Scientific Research
ISBN 0-915928-82-5

90. Fundamentals of Solid-Propellant Combustion (1984)
Kenneth K. Kuo
Pennsylvania State University
Martin Summerfield
Princeton Combustion Research Laboratories, Inc.
ISBN 0-915928-84-1

91. Spacecraft Contamination: Sources and Prevention (1984)
J. A. Roux
University of Mississippi
T. D. McCay
NASA Marshall Space Flight Center
ISBN 0-915928-85-X

92. Combustion Diagnostics by Nonintrusive Methods (1984)
T. D. McCay
NASA Marshall Space Flight Center
J. A. Roux
University of Mississippi
ISBN 0-915928-86-8

93. The INTELSAT Global Satellite System (1984)
Joel Alper
COMSAT Corp.
Joseph Pelton
INTELSAT
ISBN 0-915928-90-6

94. Dynamics of Shock Waves, Explosions, and Detonations (1984)
J. R. Bowen
University of Washington
N. Manson
Universite de Poitiers
A. K. Oppenheim
University of California at Berkeley
R. I. Soloukhin
Institute of Heat and Mass Transfer, BSSR Academy of Sciences
ISBN 0-915928-91-4

95. Dynamics of Flames and Reactive Systems (1984)
J. R. Bowen
University of Washington
N. Manson
Universite de Poitiers
A. K. Oppenheim
University of California at Berkeley
R. I. Soloukhin
Institute of Heat and Mass Transfer, BSSR Academy of Sciences
ISBN 0-915928-92-2

96. Thermal Design of Aeroassisted Orbital Transfer Vehicles (1985)
H. F. Nelson
University of Missouri-Rolla
ISBN 0-915928-94-9

97. Monitoring Earth's Ocean, Land, and Atmosphere from Space—Sensors, Systems, and Applications (1985)
Abraham Schnapf
Aerospace Systems Engineering
ISBN 0-915928-98-1

---
*Out of print.

98. Thrust and Drag: Its Prediction and Verification (1985)
Eugene E. Covert
Massachusetts Institute of Technology
C. R. James
Vought Corp.
William F. Kimzey
Sverdrup Technology AEDC Group
George K. Richey
U.S. Air Force
Eugene C. Rooney
U.S. Navy Department of Defense
ISBN 0-930403-00-2

99. Space Stations and Space Platforms—Concepts, Design, Infrastructure, and Uses (1985)
Ivan Bekey
Daniel Herman
NASA Headquarters
ISBN 0-930403-01-0

100. Single- and Multi-Phase Flows in an Electromagnetic Field: Energy, Metallurgical, and Solar Applications (1985)
Herman Branover
Ben-Gurion University of the Negev
Paul S. Lykoudis
Purdue University
Michael Mond
Ben-Gurion University of the Negev
ISBN 0-930403-04-5

101. MHD Energy Conversion: Physiotechnical Problems (1986)
V. A. Kirillin
A. E. Sheyndlin
Soviet Academy of Sciences
ISBN 0-930403-05-3

102. Numerical Methods for Engine-Airframe Integration (1986)
S. N. B. Murthy
Purdue University
Gerald C. Paynter
Boeing Airplane Co.
ISBN 0-930403-09-6

103. Thermophysical Aspects of Re-Entry Flows (1986)
James N. Moss
NASA Langley Research Center
Carl D. Scott
NASA Johnson Space Center
ISBN 0-930430-10-X

***104. Tactical Missile Aerodynamics (1986)**
M. J. Hemsch
PRC Kentron, Inc.
J. N. Nielson
NASA Ames Research Center
ISBN 0-930403-13-4

105. Dynamics of Reactive Systems Part I: Flames and Configurations; Part II: Modeling and Heterogeneous Combustion (1986)
J. R. Bowen
University of Washington
J.-C. Leyer
Universite de Poitiers
R. I. Soloukhin
Institute of Heat and Mass Transfer, BSSR Academy of Sciences
ISBN 0-930403-14-2

106. Dynamics of Explosions (1986)
J. R. Bowen
University of Washington
J.-C. Leyer
Universite de Poitiers
R. I. Soloukhin
Institute of Heat and Mass Transfer, BSSR Academy of Sciences
ISBN 0-930403-15-0

***107. Spacecraft Dielectric Material Properties and Spacecraft Charging (1986)**
A. R. Frederickson
U.S. Air Force Rome Air Development Center
D. B. Cotts
SRI International
J. A. Wall
U.S. Air Force Rome Air Development Center
F. L. Bouquet
Jet Propulsion Laboratory, California Institute of Technology
ISBN 0-930403-17-7

***108. Opportunities for Academic Research in a Low-Gravity Environment (1986)**
George A. Hazelrigg
National Science Foundation
Joseph M. Reynolds
Louisiana State University
ISBN 0-930403-18-5

109. Gun Propulsion Technology (1988)
Ludwig Stiefel
U.S. Army Armament Research, Development and Engineering Center
ISBN 0-930403-20-7

*Out of print.

110. Commercial Opportunities in Space (1988)
F. Shahrokhi
K. E. Harwell
University of Tennessee Space Institute
C. C. Chao
National Cheng Kung University
ISBN 0-930403-39-8

111. Liquid-Metal Flows: Magnetohydrodynamics and Application (1988)
Herman Branover
Michael Mond
Yeshajahu Unger
Ben-Gurion University of the Negev
ISBN 0-930403-43-6

112. Current Trends in Turbulence Research (1988)
Herman Branover
Micheal Mond
Yeshajahu Unger
Ben-Gurion University of the Negev
ISBN 0-930403-44-4

113. Dynamics of Reactive Systems Part I: Flames; Part II: Heterogeneous Combustion and Applications (1988)
A. L. Kuhl
R&D Associates
J. R. Bowen
University of Washington
J.-C. Leyer
Universite de Poitiers
A. Borisov
USSR Academy of Sciences
ISBN 0-930403-46-0

114. Dynamics of Explosions (1988)
A. L. Kuhl
R & D Associates
J. R. Bowen
University of Washington
J.-C. Leyer
Universite de Poitiers
A. Borisov
USSR Academy of Sciences
ISBN 0-930403-47-9

115. Machine Intelligence and Autonomy for Aerospace (1988)
E. Heer
Heer Associates, Inc.
H. Lum
NASA Ames Research Center
ISBN 0-930403-48-7

116. Rarefied Gas Dynamics: Space Related Studies (1989)
E. P. Muntz
University of Southern California
D. P. Weaver
U.S. Air Force Astronautics Laboratory (AFSC)
D. H. Campbell
University of Dayton Research Institute
ISBN 0-930403-53-3

117. Rarefied Gas Dynamics: Physical Phenomena (1989)
E. P. Muntz
University of Southern California
D. P. Weaver
U.S. Air Force Astronautics Laboratory (AFSC)
D. H. Campbell
University of Dayton Research Institute
ISBN 0-930403-54-1

118. Rarefied Gas Dynamics: Theoretical and Computational Techniques (1989)
E. P. Muntz
University of Southern California
D. P. Weaver
U.S. Air Force Astronautics Laboratory (AFSC)
D. H. Campbell
University of Dayton Research Institute
ISBN 0-930403-55-X

119. Test and Evaluation of the Tactical Missile (1989)
Emil J. Eichblatt Jr.
Pacific Missile Test Center
ISBN 0-930403-56-8

120. Unsteady Transonic Aerodynamics (1989)
David Nixon
Nielsen Engineering & Research, Inc.
ISBN 0-930403-52-5

121. Orbital Debris from Upper-Stage Breakup (1989)
Joseph P. Loftus Jr.
NASA Johnson Space Center
ISBN 0-930403-58-4

122. Thermal-Hydraulics for Space Power, Propulsion and Thermal Management System Design (1990)
William J. Krotiuk
General Electric Co.
ISBN 0-930403-64-9

*Out of print.

123. Viscous Drag Reduction in Boundary Layers (1990)
Dennis M. Bushnell
Jerry N. Hefner
NASA Langley Research Center
ISBN 0-930403-66-5

***124. Tactical and Strategic Missile Guidance (1990)**
Paul Zarchan
Charles Stark Draper Laboratory, Inc.
ISBN 0-930403-68-1

125. Applied Computational Aerodynamics (1990)
P. A. Henne
Douglas Aircraft Company
ISBN 0-930403-69-X

126. Space Commercialization: Launch Vehicles and Programs (1990)
F. Shahrokhi
University of Tennessee Space Institute
J. S. Greenberg
Princeton Synergetics Inc.
T. Al-Saud
Ministry of Defense and Aviation Kingdom of Saudi Arabia
ISBN 0-930403-75-4

127. Space Commercialization: Platforms and Processing (1990)
F. Shahrokhi
University of Tennessee Space Institute
G. Hazelrigg
National Science Foundation
R. Bayuzick
Vanderbilt University
ISBN 0-930403-76-2

128. Space Commercialization: Satellite Technology (1990)
F. Shahrokhi
University of Tennessee Space Institute
N. Jasentuliyana
United Nations
N. Tarabzouni
King Abulaziz City for Science and Technology
ISBN 0-930403-77-0

129. Mechanics and Control of Large Flexible Structures (1990)
John L. Junkins
Texas A&M University
ISBN 0-930403-73-8

130. Low-Gravity Fluid Dynamics and Transport Phenomena (1990)
Jean N. Koster
Robert L. Sani
University of Colorado at Boulder
ISBN 0-930403-74-6

131. Dynamics of Deflagrations and Reactive Systems: Flames (1991)
A. L. Kuhl
Lawrence Livermore National Laboratory
J.-C. Leyer
Universite de Poitiers
A. A. Borisov
USSR Academy of Sciences
W. A. Sirignano
University of California
ISBN 0-930403-95-9

132. Dynamics of Deflagrations and Reactive Systems: Heterogeneous Combustion (1991)
A. L. Kuhl
Lawrence Livermore National Laboratory
J.-C. Leyer
Universite de Poitiers
A. A. Borisov
USSR Academy of Sciences
W. A. Sirignano
University of California
ISBN 0-930403-96-7

133. Dynamics of Detonations and Explosions: Detonations (1991)
A. L. Kuhl
Lawrence Livermore National Laboratory
J.-C. Leyer
Universite de Poitiers
A. A. Borisov
USSR Academy of Sciences
W. A. Sirignano
University of California
ISBN 0-930403-97-5

134. Dynamics of Detonations and Explosions: Explosion Phenomena (1991)
A. L. Kuhl
Lawrence Livermore National Laboratory
J.-C. Leyer
Universite de Poitiers
A. A. Borisov
USSR Academy of Sciences
W. A. Sirignano
University of California
ISBN 0-930403-98-3

*Out of print.

135. Numerical Approaches to Combustion Modeling (1991)
Elaine S. Oran
Jay P. Boris
Naval Research Laboratory
ISBN 1-56347-004-7

136. Aerospace Software Engineering (1991)
Christine Anderson
U.S. Air Force Wright Laboratory
Merlin Dorfman
Lockheed Missiles & Space Company, Inc.
ISBN 1-56347-005-0

137. High-Speed Flight Propulsion Systems (1991)
S. N. B. Murthy
Purdue University
E. T. Curran
Wright Laboratory
ISBN 1-56347-011-X

138. Propagation of Intensive Laser Radiation in Clouds (1992)
O. A. Volkovitsky
Yu. S. Sedenov
L. P. Semenov
Institute of Experimental Meteorology
ISBN 1-56347-020-9

139. Gun Muzzle Blast and Flash (1992)
Günter Klingenberg
Fraunhofer-Institut für Kurzzeitdynamik, Ernst-Mach-Institut
Joseph M. Heimerl
U.S. Army Ballistic Research Laboratory
ISBN 1-56347-012-8

140. Thermal Structures and Materials for High-Speed Flight (1992)
Earl. A. Thornton
University of Virginia
ISBN 1-56347-017-9

141. Tactical Missile Aerodynamics: General Topics (1992)
Michael J. Hemsch
Lockheed Engineering & Sciences Company
ISBN 1-56347-015-2

142. Tactical Missile Aerodynamics: Prediction Methodology (1992)
Michael R. Mendenhall
Nielsen Engineering & Research, Inc.
ISBN 1-56347-016-0

143. Nonsteady Burning and Combustion Stability of Solid Propellants (1992)
Luigi De Luca
Politecnico di Milano
Edward W. Price
Georgia Institute of Technology
Martin Summerfield
Princeton Combustion Research Laboratories, Inc.
ISBN 1-56347-014-4

144. Space Economics (1992)
Joel S. Greenberg
Princeton Synergetics, Inc.
Henry R. Hertzfeld
HRH Associates
ISBN 1-56347-042-X

145. Mars: Past, Present, and Future (1992)
E. Brian Pritchard
NASA Langley Research Center
ISBN 1-56347-043-8

146. Computational Nonlinear Mechanics in Aerospace Engineering (1992)
Satya N. Atluri
Georgia Institute of Technology
ISBN 1-56347-044-6

147. Modern Engineering for Design of Liquid-Propellant Rocket Engines (1992)
Dieter K. Huzel
David H. Huang
Rocketdyne Division of Rockwell International
ISBN 1-56347-013-6

148. Metallurgical Technologies, Energy Conversion, and Magnetohydrodynamic Flows (1993)
Herman Branover
Yeshajahu Unger
Ben-Gurion University of the Negev
ISBN 1-56347-019-5

149. Advances in Turbulence Studies (1993)
Herman Branover
Yeshajahu Unger
Ben-Gurion University of the Negev
ISBN 1-56347-018-7

150. Structural Optimization: Status and Promise (1993)
Manohar P. Kamat
Georgia Institute of Technology
ISBN 1-56347-056-X

---
*Out of print.

151. Dynamics of Gaseous Combustion (1993)
A. L. Kuhl
Lawrence Livermore National Laboratory
J.-C. Leyer
Universite de Poitiers
A. A. Borisov
USSR Academy of Sciences
W. A. Sirignano
University of California
ISBN 1-56347-060-8

152. Dynamics of Heterogeneous Gaseous Combustion and Reacting Systems (1993)
A. L. Kuhl
Lawrence Livermore National Laboratory
J.-C. Leyer
Universite de Poitiers
A. A. Borisov
USSR Academy of Sciences
W. A. Sirignano
University of California
ISBN 1-56347-058-6

153. Dynamic Aspects of Detonations (1993)
A. L. Kuhl
Lawrence Livermore National Laboratory
J.-C. Leyer
Universite de Poitiers
A. A. Borisov
USSR Academy of Sciences
W. A. Sirignano
University of California
ISBN 1-56347-057-8

154. Dynamic Aspects of Explosion Phenomena (1993)
A. L. Kuhl
Lawrence Livermore National Laboratory
J.-C. Leyer
Universite de Poitiers
A. A. Borisov
USSR Academy of Sciences
W. A. Sirignano
University of California
ISBN 1-56347-059-4

155. Tactical Missile Warheads (1993)
Joseph Carleone
Aerojet General Corporation
ISBN 1-56347-067-5

156. Toward a Science of Command, Control, and Communications (1993)
Carl R. Jones
Naval Postgraduate School
ISBN 1-56347-068-3

157. Tactical and Strategic Missile Guidance Second Edition (1994)
Paul Zarchan
Charles Stark Draper Laboratory, Inc.
ISBN 1-56347-077-2

158. Rarefied Gas Dynamics: Experimental Techniques and Physical Systems (1994)
Bernie D. Shizgal
University of British Columbia
David P. Weaver
Phillips Laboratory
ISBN 1-56347-079-9

159. Rarefied Gas Dynamics: Theory and Simulations (1994)
Bernie D. Shizgal
University of British Columbia
David P. Weaver
Phillips Laboratory
ISBN 1-56347-080-2

160. Rarefied Gas Dynamics: Space Sciences and Engineering (1994)
Bernie D. Shizgal
University of British Columbia
David P. Weaver
Phillips Laboratory
ISBN 1-56347-081-0

161. Teleoperation and Robotics in Space (1994)
Steven B. Skaar
University of Notre Dame
Carl F. Ruoff
Jet Propulsion Laboratory, California Institute of Technology
ISBN 1-56347-095-0

162. Progress in Turbulence Research (1994)
Herman Branover
Yeshajahu Unger
Ben-Gurion University of the Negev
ISBN 1-56347-099-3

163. Global Positioning System: Theory and Applications, Volume I (1996)
Bradford W. Parkinson
Stanford University
James J. Spilker Jr.
Stanford Telecom
Penina Axelrad, Associate Editor
University of Colorado
Per Enge, Associate Editor
Stanford University
ISBN 1-56347-107-8

164. Global Positioning System: Theory and Applications, Volume II (1996)
Bradford W. Parkinson
Stanford University
James J. Spilker Jr.
Stanford Telecom
Penina Axelrad, Associate Editor
University of Colorado
Per Enge, Associate Editor
Stanford University
ISBN 1-56347-106-X

---

*Out of print.

165. Developments in High-Speed Vehicle Propulsion Systems (1996)
S. N. B. Murthy
Purdue University
E. T. Curran
Wright Laboratory
ISBN 1-56347-176-0

166. Recent Advances in Spray Combustion: Spray Atomization and Drop Burning Phenomena, Volume I (1996)
Kenneth K. Kuo
Pennsylvania State University
ISBN 1-56347-175-2

167. Fusion Energy in Space Propulsion (1995)
Terry Kammash
University of Michigan
ISBN 1-56347-184-1

168. Aerospace Thermal Structures and Materials for a New Era (1995)
Earl A. Thornton
University of Virginia
ISBN 1-56347-182-5

169. Liquid Rocket Engine Combustion Instability (1995)
Vigor Yang
William E. Anderson
Pennsylvania State University
ISBN 1-56347-183-3

170. Tactical Missile Propulsion (1996)
G. E. Jensen
United Technologies Corporation
David W. Netzer
Naval Postgraduate School
ISBN 1-56347-118-3

171. Recent Advances in Spray Combustion: Spray Combustion Measurements and Model Simulation, Volume II (1996)
Kenneth K. Kuo
Pennsylvania State University
ISBN 1-56347-181-7

172. Future Aeronautical and Space Systems (1997)
Ahmed K. Noor
NASA Langley Research Center
Samuel L. Venneri
NASA Headquarters
ISBN 1-56347-188-4

173. Advances in Combustion Science: In Honor of Ya. B. Zel'dovich (1997)
William A. Sirignano
University of California
Alexander G. Merzhanov
Russian Academy of Sciences
Luigi De Luca
Politecnico di Milano
ISBN 1-56347-178-7

174. Fundamentals of High Accuracy Inertial Navigation (1997)
Averil B. Chatfield
ISBN 1-56347-243-0

175. Liquid Propellant Gun Technology (1997)
Günter Klingenberg
Fraunhofer-Institut für Kurzzeitdynamik, Ernst-Mach-Institut
John D. Knapton
Walter F. Morrison
Gloria P. Wren
U.S. Army Research Laboratory
ISBN 1-56347-196-5

176. Tactical and Strategic Missile Guidance Third Edition (1998)
Paul Zarchan
Charles Stark Draper Laboratory, Inc.
ISBN 1-56347-279-1

177. Orbital and Celestial Mechanics (1998)
John P. Vinti
Gim J. Der, Editor
TRW
Nino L. Bonavito, Editor
NASA Goddard Space Flight Center
ISBN 1-56347-256-2

178. Some Engineering Applications in Random Vibrations and Random Structures (1998)
Giora Maymon
RAFAEL
ISBN 1-56347-258-9

179. Conventional Warhead Systems Physics and Engineering Design (1998)
Richard M. Lloyd
Raytheon Systems Company
ISBN 1-56347-255-4

180. Advances in Missile Guidance Theory (1998)
Joseph Z. Ben-Asher
Isaac Yaesh
Israel Military Industries—Advanced Systems Division
ISBN 1-56347-275-9

*Out of print.